Mobile
Communication

Mobile
Communication

Dimensions of Social Policy

James E. Katz
editor

 Routledge
Taylor & Francis Group

LONDON AND NEW YORK

First published 2011 by Transaction Publishers

2 Park Square, Milton Park, Abingdon, Oxfordshire OX14 4RN
711 Third Avenue, New York, NY 10017

Routledge is an imprint of the Taylor & Francis Group, an informa business

First issued in paperback 2017

Library of Congress Catalog Number: 2010035053

Library of Congress Cataloging-in-Publication Data

Mobile communication : dimensions of social policy / [edited by] James E. Katz.
     p. cm.
   Includes index.
   ISBN 978-1-4128-1468-3 (alk. paper)
   1. Cell phones—Social aspects. 2. Mobile communication—Social aspects.
3. Social policy. I. Katz, James Everett.
   HE9713.M6327 2011
   303.48'33—dc22
                    2010035053

ISBN 13: 978-1-4128-1468-3 (hbk)
ISBN 13: 978-1-138-51227-6 (pbk)

To

Rachel, Raymond, and Franchesca

For ye shall go out with joy, and be led forth with peace;
the mountains and the hills shall break forth before you into
singing, and all the trees of the field shall clap their hands.

Isaiah 55:12

CONTENTS

Preface and Acknowledgments

The Era of Mobile Communication

It is apparent to anyone who walks along a busy street nowadays or, for that matter, to anyone who tries to cross one, that an object of human desire is to communicate to distant others, and to do so while being mobile. The mobile device is a commanding presence in the daily lives of many if not most people. With little exaggeration, the first part of the twenty-first century could be characterized as the era of mobile communication. The *Zeitgeist* of mobiles may be seen in their phenomenal success. More than 5 billion of the world's 6.5 billion people are now mobile users. Among their number are not only many children but also those, such as the Amish, who for religious reasons have been forbidden to have earlier forms of the telephone. Also included in this number are those who have been forbidden by public policy to have access to mobile phones, such as prisoners. Indeed, incarcerated gang members in Brazil have used their illicit mobile phones to organize simultaneous riots in prisons across the country.

Yet the numerical aspects of the success of mobile communication tell only part of the tale. Another aspect is experiential: one sees incessant usage in practically all venues. Long gone are the days when it would seem odd for a couple to walk along the street together while each talked on their respective mobiles to some absent other. Yet the interpersonal difficulties of juggling absent others with present selves are still with us, causing at least some adverse comments. Such incidents include a major political address. During one such event in 2008, the speaker, US presidential aspirant Rudy Giuliani, interrupted himself to answer his ringing mobile phone. (It turned out to be his wife calling; he denied that the call was a planned stunt.) They also include official functions. Such an incident occurred during a 2009 NATO meeting when arriving Italian Prime Minister Silvio Berlusconi left his host, German Chancellor Angela Merkel, standing by herself while he walked off to take a phone call. But politicians and dignitaries are not alone in such behavior. Mobiles are used in hospital rooms as a child is being brought into the world and during graveside services when a loved one is being escorted out. In fact, people's involvement with the mobile phone now stretches not only from cradle to grave, but into the grave itself as people do not infrequently ask to be buried with their mobile phones.

The mobile is not only omnipresent, but seems omnivorous, devouring coextensive objects of ordinary life. It incorporates quotidian objects such as cameras, wristwatches, calendars, landline phones, public phone booths, and even the TV. The mobile is beginning to supersede the wallet, credit card, ID card, and medical dossier. The mobile is not only challenging bankcards, but the bank itself as it becomes an interface to an independent international financial system. The mobile conjoins two important attributes: a communication network and a financial network, and does so at levels that can address the needs of people from both industrialized and hunter-gatherer societies, with tasks ranging from global stock trading to dispensing a child's allowance. A common invisible thread now binds a Wall Street financer who upon emplaning a Singapore-bound jet uses her mobile phone as a boarding pass with a farmer in Kakamega, Kenya who receives "gift credit" of additional minutes to his mobile phone account from a cousin in Nairobi.

Scholars have been busily figuring out what the reality of mobile technology is doing to people and how people are creating and re-creating uses for mobile devices and services. Much of this inquiry has been devoted to mobile communication's sociological, psychological and interactional aspects. The role of the mobile in economic development, especially in poor areas, has been subjected to numerous studies. The cultural studies movement has the mobile firmly fixed in its analytical gaze. All these endeavors have allowed sharper understandings of the mobile communication era. That said, it is also true that comparatively less attention has been directed to the policy ramifications of mobile communication technology. The present volume is an attempt to help stimulate greater scholarly attention to the subject and grows out of an October 2009 conference organized by the Center for Mobile Communication Studies supported by a grant from the Horowitz Foundation for Social Policy.

The volume by no means represents a definitive statement on the subject of mobile communication social policy: in fact, given the dimensions of the mobile communication phenomenon, there could be no such book. Rather the intention here is provide samples and perspectives to show the rich array of possibilities in the area. The book's aim is to open further, rather than cap off, scholarly investigation of a fertile topic.

Acknowledgments

Holding a conference or editing a book inevitably incurs many obligations. This section provides me with a welcome opportunity to thank some of the many colleagues and students who have been instrumental to the success of this undertaking.

For the book itself, which grew out of an event sponsored by the Horowitz Foundation for Social Policy, I express my gratitude to Irving Louis Horowitz and Mary Curtis. Their pivotal roles at Transaction Publishers have done much to foster studies of mobile communication, by launching a series of books in mobile communication at Transaction. Of course this project could not have been possible, or at least would have taken longer and been of reduced quality, had not the Horowitz Foundation for Social Policy sponsored the conference.

With his usual verve and ever-inquiring mind, Irving Louis Horowitz grasped adroitly the gap between technological change, especially in terms of mobile communication, and the moral and policy regimes within which such changes occur. Horowitz, a scholar of the first order, was my dissertation advisor in the early 1970s. Since that time, he has been a friend, colleague and inspiration. He encouraged me to organize a conference at Rutgers in order to spark greater scholarly attention to the intersection of mobile communication technology and social policy analysis. Such a conference would meld several of his interests that pivot around not only policy per se but also the exercise of freedom and the use of power at every level of society. His role as chairman of the Horowitz Foundation for Social Policy complemented mine as director of the Rutgers University Center for Mobile Communication Studies. With funding from his foundation, my center was able in October 2009 to hold a conference at Rutgers. Scholars came from nearly every continent to examine the area of mobile communication and social policy. At the conference, we were able to discuss overarching issues and explore common themes and novel perspectives. A selection of those papers forms the heart of the volume.

The model for an event such as this grew out of a vision of colleague Hartmut B. Mokros. With his encouragement, Mark Aakhus and I in 1999 had the good fortune to organize what turned out to be a seminal workshop at Rutgers entitled *Perpetual Contact*. That workshop was apparently the first international social scientific event to examine how mobile communication was changing everyday life; it was certainly a turning point in the careers for many of the scholars who attended. Harty

Mokros' acute intellectual insight and personal commitment to inspiring and guiding other scholars was made manifest in this instance. Yet this was no exception, rather it is part and parcel of his daily life: he works to bring out the best in other scholars even as he continues to develop meaningful insights through his own research.

Several leading scholars have done much to illuminate the field for me. Manuel Castells has been a canny and astute observer of the mobile scene for many years. Drawing on his deep analysis of networks, political economy, technological innovation and even human neurobiology, he has formulated many keen insights about the mobile network society. Kenneth J. Gergen has for years challenged me to think through the important implications of mobile communication for the quality of people's social relationships. Ken embodies the soul of a scholar and the sensibilities of a humanist. Ron Rice has continued to share his research and analysis generously with me. His collegiality and professionalism have been without equal, and his acerbic humor has lightened many a dark situation.

Turning now to those who have been the sinews that connected the world of ideas to a sense of place, I thank Chih-Hui Lai and Amanda McGarry who were instrumental in helping me organize the conference. Hui-Min Kuo deserves special mention as she anticipated and solved nearly every problem that one encounters when holding an international conference, and did so with grace and sympathy. Attendees at the conference were fortunate to encounter a variety of helpers who did much to create a productive atmosphere. These included Sun-Kyung "Sunny" Lee, Soo-yeon Hwang, Marianne Stewart-Titus, Die fie "Steven" Xie, Yumi Jung, Heewon Kim and Hyunsook Youn. Frank Bridges not only helped with the event itself, but was instrumental in helping create exciting graphics for the event. Yi-Fan Chen with quiet efficiency laid the groundwork for the event from initial organization and graphics through to online structures and updates. After the conference, Lora Appel helped organize materials and interfaced with contributors.

Rutgers colleagues played an important part in adding to the success of the event and have, at a more general level, been helpful colleagues as I have sought to develop mobile communication studies at Rutgers. These colleagues include Jorge Reina Schement, Michael Pazzani and Brent Ruben, all of whom offered sagacious comments at the conference's welcoming dinner. Colleagues in the Department of Communication and who are also members of the Center for Mobile Communication Studies have helped create a meaningful domicile for mobile communication studies at Rutgers; these include Craig R. Scott, Jeff Boase, Mark Aakhus and Jenn Gibbs. Within the Department's home, the Rutgers School of Communication and Information, I have benefited from the support and advice of Karen Novick, Michael Lesk,Claire McInerney, John V. Pavlik and Richard D. Heffner. From the larger Rutgers family, I would like to highlight the ongoing encouragement and collegial support I have received in this endeavor and many others from my colleagues Myoung Wilson, Douglas Greenberg, Ed Tenner and Lionel Tiger.

Looking beyond the banks of the Raritan River, above which is perched Rutgers, I am indebted to North American colleagues John L. King, Scott Campbell, Ran Wei, Katie Lever-Mazzuto, Gary Kreps, John Carey, Naomi Baron, Amit Schejter, Nojin Kwak, Barry Wellman, and Sharon Strover.

Looking still further from the Raritan, across the seas, I would like to thank colleagues from beyond North America who also played a major role in this project. These include Leopoldina Fortunati, Rich Ling, Joachim Höflich, Kristóf Nyíri, Marinella Rocca Longo, Boxu Yang, Gerard Goggin, Akiba Cohen, Jonathan Donner, Matteo Tarantino, Pui-lam "Patrick" Law, Ilpo Koskinen, Satomi Sugiyama, Christian Licoppe, Peppino Ortoleva, Mirjam deBrujin, Andre Caron, Letizia Caronia, Guiseppina Pellegrino, Kirsten Drotner, Imar deVries, Gordon Gow, Nello Barile, and Raul Pertierra.

To these fine people, I offer my heartfelt thanks.

James E. Katz
New Brunswick, New Jersey

1

Editor's Introduction

James E. Katz

The mobile device has changed daily life in so many ways that only its absence reminds us of how dependent we have become on it. When there is a mobile network outage, as happened to T-Mobile users in November 2009, users experience an often painful and disturbing return to life before the mobile. Though some found Zen-like escape, angst and anger were more typical. For many, the economic losses were substantial. Nowadays, not having a mobile is a form of social death.

Yet on the other side, we are only at the dawn of a world of mobile applications. These are quickly becoming part of the way social policy is influenced and delivered. As well, policy for mobile communication is also a form of social policy.

A mere sampling of topics demonstrates the enormous breadth of the field, as may be seen from the following list of salient issues, which itself is quite incomplete:

- Identity and anonymity in location and communication, including advertisements and sharing of personal information in profiles
- Access to and delivery of social services via mobiles
- Mobile banking, mobile money and its commercial and economic development consequences
- Mobile communication and economic development, especially in peripheral or remote areas
- Communication with and responsiveness of political leaders to mobile-equipped citizens
- Role of mobile communication in political activism and organizing
- Policy-making for mobiles, including spectrum allocation, user safety and traceability of usage
- Policy for use of mobiles while driving, flying or engaging in other activities
- Regulation of mobile use in cultural institutions and other public places

- First responders/crisis management/emergency notification/disaster recovery
- Issues of national integration and diasporas related to use of mobile communication
- Mobile learning, both free choice and structured, as well as implications for the classroom
- Wellness, health and the role of the mobile in issues of health care provider responsiveness, liability and access; individual health monitoring
- Mobile communication in public health campaigns and other social engineering efforts
- Mobile device accessibility and integration/empowerment of handicapped people: equal opportunity and limitations of affordances of mobile devices
- Mobile communication and information overload, social relationships, "always on" mentality and psychological stress
- Mobile phone/device crimes and crime prevention: effect on deployment of police resources as well as monitoring of police units; summoning help and reporting crime
- Religious and spiritual activities: their implications for social practices, evangelism, and institutional control
- Interpersonal contact services and related commercial activities: their implications for stalking, privacy, anonymity, and the development of close personal relationships
- Use of mobile devices worn on the body (for house arrest or GPS criminal offender supervision)
- Health status monitoring, including insertion into the body of various locational reporting devices

Not all of these issues will be tackled in the present volume, far from it. Rather, a selection of topics will be examined in the book with the intention of signaling possibilities and exploring specific topics and situations in their incipiency. This chapter provides a brief overview of the contents of the book.

In a chapter designed to highlight topics, Lora Appel, Amanda McGarry and I present an analysis of some salient issues related to the topic of mobile communication social policy. We suggest a few issues that are of compelling interest to scholars and researchers and examine them. These include questions of parent-child social relations, the rise of mobile financial transactions, anonymity and confidentiality, and voting.

John L. King offers a socio-technical perspective on mobile communication as product of industrial policy. He emphasizes the importance of infrastructure, which tends to disappear when one is examining a particular technology. With this intriguing departure point, he examines the interplay between ideas, routines, and technological progress.

By examining evolutionary communication technologies, he is able to formulate acute insights into the culture of expectations and policy and the way decisions affect the micro-structures of daily interaction, both historically for telephony generally and most especially in the case of mobile communication technology.

Gordon Gow and Nuwan Waidyanatha discuss the challenges of sustainable early warning systems. They explore how mobile phones can provide early warning systems, especially in terms of the so-called last-mile problem, that is, getting the warning information to the very people whose lives are at stake. To examine this issue, they look at Sri Lanka. Their research shows that the mobile phone can be more easily integrated into everyday life as opposed to more specialized warning technologies. They find out that, by itself, access is not sufficient to sustain an adequate level of emergency preparedness at the level of local communities. Their review of the literature shows that local risk knowledge is an important component to developing effective community-based strategies. They offer the intriguing idea that effectiveness could be fostered by engaging in a process of ritual interaction that establishes the value of local risk knowledge. They believe their prescription would be a cost-effective way to create sustainable disaster preparedness and early warning capabilities.

Rich Ling and Nisar Bashir look at the promise of mobile technology as part of the worldwide move towards environmentally protective ("green") technology. They point out that mobile communication can help people use energy more efficiently and reduce their detrimental impact on the environment. Nonetheless, they point out that mobile communication itself has various environmental consequences that make the situation more complex than might first be appreciated. For instance, on the one hand, mobile communication can help traffic flow more efficiently and can reduce material waste, as for example moving pulp-based reading material like newspapers on to mobile devices. On the other hand, mobile phones themselves require careful thought when it comes to their ultimate disposal. Many of the substances in them are dangerous if not removed from the environment. In fact, it is estimated that several hundred million mobile phones are stored in drawers in homes around the world and at some point will have to be disposed of properly or else they could pose an environmental risk. These are but a few of the host of environmental issues in which the mobile phone plays an important role in both ameliorating and exacerbating the situation. They conclude that systemic policy responses are in order.

In his chapter, Yun Xia presents a perspective on the crisis response to the earthquake on May 12, 2008 in Sichuan province, China. He proposes specific patterns that are identifiable from the ethnographic fieldwork. First, the Chinese used their mobile phones to share information and reduce uncertainty. They sought out their friends and relatives at a time of crisis. (This parallels the work of Akiba Cohen and others.) He notes how the limits of ordinary terrestrial systems can be highlighted during crises such as an earthquake. This results in acceleration of programs to create satellite-based systems. At the same time, he shows how text-messaging enabled people to bypass official communication channels and learn for themselves what the real situation was. This of course poses a serious challenge to the official media. It also prompts either a crackdown on unofficial person-to-person communication (what Mao Tse-tung described as "small broadcasts") or a more robust and responsive media at the official level.

Scott W. Campbell and Nojin Kwak explore the use of mobile communication, social networks and policy knowledge during the 2008 US presidential election. To do this, they explore the concept that Gergen proposed in 2008 of a "monadic cluster." A monadic cluster is one in which small enclaves of like-minded individuals work closely together. Although such ties bind one group together, they could also impede public discussion as well as democratic participation. The authors seek to get an empirical grasp on this question by surveying individuals with strong network ties and seeing to what extent they have knowledge of the policy position of candidates. They find evidence that the increased use of technology in small networks and among those who are like-minded did lead to reduced policy knowledge. In contrast, those with extensive mobile-mediated discussion in a politically diverse network had greater knowledge. They conclude by discussing policy implications of this work.

In their chapter on South Africa's 2009 elections, Marion Walton and Jonathan Donner discuss four kinds of mobile media and political participation. They find SMS "wars" in the time leading up to the election, and see that these are part of open contention over political issues. They also find ".mobi" websites, which are hosted by political parties, offer a great deal of mobile-generated online content. They discover that the content thus generated, and its uses, failed to ignite the hoped-for engagement of the public. This highlight of how the mainstream political parties sought to energize their campaigns with the use of mobiles and other technology is similar to that which characterized Barack

Obama's 2008 campaign in the United States. These initiatives were not successful. By contrast, the African National Congress' arm, The Youth League, was able to adopt effective mobile social network campaigns. The authors highlight the fact that mobile communication can actually raise barriers and obstacles to participation, despite the well-advertised claims that they enhanced political participation. Thus their comments are another cautionary tale to the temptation to overly romanticize the possibility of new technology to engage public participation.

Rhonda McEwen and Melissa Fritz investigate mobile phone practices and social policy related to electromagnetic frequencies (EMF). They conducted numerous interviews with Canadian policy makers. They examined the explanations offered by these policymakers as to why no directives have been developed concerning EMF. This is a particular concern because of the ever-growing exposure of children and youth to such energy sources. This is all the more remarkable because other G8 countries have been investigating the problem rather aggressively. The authors pose a novel explanation, namely that policymakers do not understand youth mobile phone practices. When combined with a lack of understanding of inadequate policy framework with which to deal with the problem, this has led to little or no progress in the policy domain. The chapter is relevant not only for its empirical exploration through interviews, but also its use of discourse analysis to highlight the problem space.

Craig R. Scott, Hyunsook Youn, and Gillian Bonanno begin their contribution by noting that mobiles are increasingly ubiquitous in the workplace and, consequently, organizational policies governing them have proliferated. Despite their pervasiveness in the work environment, and the importance of rules governing their use, they note that little scholarly attention has been given to the analysis of such policies. To address this gap, the authors analyze policies towards mobile communication from state governments in the United States. They identify 15 themes related to some aspect of mobile communication and reduce these themes to four categories: supplemental/background, monitoring/tracking, eligibility, and user/use. Perhaps not surprisingly, they find wide variation in specific policies. They find that there can be a great deal of ambiguity and policy language that both can help and impede rational use. They also note that language often seems to be directed towards several different, and often contending, stakeholders. This analysis turns up many intriguing implications for both the formal analysis of communication strategies in organizations as well as the development of public policies that are responsive to both the personal works in the organization and the qualities that are incumbent upon a publicly funded institution.

Mina Thomas and Sun Sun Lim explore ICT use by female migrant workers in Singapore. They did this in order to better understand the received benefits and costs to ICUs among transnational migrant workers. They note that these live-in maids, most of whom are migrant workers, revolve around the lives of their employers. This isolates them physically, so ICTs become even more crucial to maintaining their social networks. Their rather expansive vision of what should be done to address this includes a training program for migrants so that they can avail themselves of ICTs. They also advocate a "Bill of Rights" between employers and workers specifying what their rights to communication technology might be, including, and especially, mobile technology.

Loudres M. Portus explores the mobile phone's contribution to, or sometimes more slyly, interference with, education. She does this by examining the use of mobile communication technology in the Philippines. She identifies many positive attributes of the technology and places it within an interesting cultural context. She highlights the fact that true learning is not merely assimilation of information, but creative application of that knowledge, as well as creative enrichment to it when appropriate.

Matteo Tarantino examines how meaning of communication is interpreted through the lens of the national media by the case study of the so-called Genchi Archive scandal. The scandal revolved around a professional interpreter who apparently amassed an archive of information on millions of Italians. In juxtaposition and contradiction of expected positions, Tarantino finds that Leftist voices defended the interceptions while all voices from the conservative perspective criticize them. He uses these findings to suggest how mobile phone technology has created a world far-different from the traditional one. That is, it is a new way to penetrate individual life-worlds and break traditional boundaries and understandings. In this way, he shows that the mobile device and its associated policies belongs not to one political party or ideology, but is reinterpreted as necessary in the service of prevailing in contested discourse.

Gerard Goggin analyzes disability and how it articulates with mobile technology. He emphasizes the fact that disability has both a physical and sociological component and then examines the role of mobile technology in facilitating the social and interactional activities of people with disabilities. He explores the potential innovative qualities of mobile communication, as well as areas in which the affordances of mobile communication have been poorly developed for people with disabilities.

Leopoldina Fortunati surveys research extending over a dozen years to show how there has been increasing social participation in five of the European countries that she studied. Nonetheless, she notes that even though social participation has increased, it is still a minority that is socially involved. She identifies mobile technology as an important pathway to effectuate greater social participation. At the same time, she is by no means a technological enthusiast. She points out that mobiles can also be used to reduce social participation. Although she comes to no firm conclusion about the long-term impact of mobile technology, she does provide an important reminder that those who look to a technological quick fix to encourage social participation will not find it any time soon in the arena of mobile technology.

Irving Louis Horowitz rounds out the volume by looking at what he calls technological rabbits and communication turtles. The title is adopted from a child's morality tale about a rabbit and a turtle that begin race. The rabbit initially runs far ahead of the turtle, but tiring and becoming distracted, ultimately loses the race to the assiduous turtle. In his essay, Horowitz compares the rapidly evolving technology to the slowly evolving moral order. However, in contrast to most analysts, he rejects the false oppositional framework of technology versus ethics. Rather he urges "public policy is best when it navigates between technology and morality." Horowitz also juxtaposes the technical analyst and the social scientist generally and the sociologist especially. He finds that many in the social science professions have overblown estimations of their expertise and importance, especially in terms of their relevance to the policy realm. Yet Horowitz also points out that the work of both these groups is necessary to steer society away from oppressive extremes of both moral relativism and technological absolutism. But before we celebrate the role of the social scientist as philosopher-mandarins, Horowitz reminds us, that hubris ends in tragedy. Hence, though we may with our professional tools aspire to do much and quickly, as did the rabbit, if we plan to win the metaphorical race it is better to remain modest in our claims.

The book concludes with my highlight of some themes that have emerged from the chapters. I also add a few of observations about future directions of mobile communication in social policy. Yet, taking a page from Horowitz's chapter, as well as a phrase used by another one of my former professors, Peter L. Berger, rather than a comprehensive examination of such a huge subject, in the closing chapter I can only offer an invitation for further reflection and study of the topic of mobile communication and social policy.

2

Overview: Surveying the Terrain

James E. Katz, Lora Appel, and Amanda McGarry

The purpose of this chapter is to sketch out some of the major questions concerning how mobile communication affects policy, and how policy is constructed for the social dimensions of mobile communication usage. These issues are so quickly unfolding, as well as vast and far-reaching, scholarly understanding of them remains incipient. Nonetheless, there are a variety of vital questions to be understood. These range from mobiles and social programs (initiation through evaluation) to how mobile technology is used in political campaigns and to influence the social policy environment.

We can explore a few of these issues to illustrate the fascinating policy questions that arise from the interaction of mobile communication and social policy. In our highly selective scan of issues, we look at the areas of (1) privacy, anonymity and publicity, (2) m-government and m-voting, (3) emergency notification, 4) parents and children with mobile phones, (5) m-education and (6) m-healthcare.

Privacy, Anonymity and Publicity

A most important policy concern, at least in terms of headlines generated in the press, as well as presumably for the long-term destiny of a free society, is the question of privacy. This concern covers both privacy of personal information about oneself and one's communication, as well as the ability to act and communicate anonymously. It also includes questions of third-party disclosure and the degree to which people's activities can be monitored, either with permission or surreptitiously. In general, it is assumed that these privacy concerns are legitimate, and the right to privacy is thus often enshrined in various laws and declarations in countries around the world (and most especially in Europe).

Concerns over privacy however must be counterbalanced with concerns over national security and public safety as well as economic efficiency. Mobile communication adds new twists to the perennial contest over privacy and anonymity versus disclosure and monitoring owing to locational information generated by the use of mobile communication devices and networks. Drawing on information generated by mobile networks, it is now possible to pinpoint the location of just about any user at any time (Harvard Law, 2004), and in many cases tap into the content of mobile communication. In terms of locational positioning, information from cell towers and networks can be triangulated to determine who is where and when. Moreover, as individual call records are kept, the calling patterns of any individual or any group can be aggregated and analyzed. This of course has many benefits such as making network resources available cost-effectively, and on a more human level, rescuing lost hikers or accident victims. A surprising array of cases has been reported ranging from rescuing someone trapped in a car trunk to locating a kidnapped child in Virginia (Barnes and Williamson, 2009). Stories abound about how the use of various mobile communication surveillance and monitoring systems have helped ordinarily people. For example, in April 2009, a California woman used Google Latitude to track down her stolen purse, which contained her Latitude-enabled Smartphone (Siegler, 2009). Still, as this particular instance demonstrates, surveillance applications are constantly accumulating and aggregating information about the users of mobile communication systems. Subscriptions to location-based service networks also leave people vulnerable to third-party marketing solicitations. For instance, when a person searches for the nearest florist on their cell phone's GPS feature, it may open them up to solicitations from other florists, in what is known as a "pull scenario" (Syme, n.d.). Again, most people are willing to reveal their location and travel habits for the convenience of immediate knowledge.

The quiet collection of user-generated information about communication patterns and location is continual. When necessary, officials can draw on these data. In this way, mobile communication records have been used to exculpate the innocent by confirming their claims of their absence from the crime scene as well as convicting the guilty by disconfirming such claims. Reaching beyond ordinary crime, reconstructed mobile phone records or even real-time interception of mobile phone calls have been instrumental in capturing or killing terrorists. These are powerful motives for the collection and analysis of mobile communication data over and above the right to privacy.

Switching analytical frames of privacy from questions of murder, mayhem, kidnapping, and terrorism are questions of ordinary life. In these terms, the location-divulging talents of the mobile have been used to provide customized advertisements, programming based on location, and even directions to the nearest gas station. There are social networking applications, such as Google Latitude, that allow users to "shout out" their location to anyone in their "friend network" and to, likewise, see their friends' locations on a map in real time. Applications such as this are designed to facilitate impromptu social gatherings, but can also encourage stalking. Similarly, Verizon Wireless offers a "Family Locator" service that allows customers to see "detailed location information and the address of your family members displayed on an interactive map" and to even get "turn by turn" directions to their family members with the assistance of Verizon's GPS application (Verizon, 2010). Once again, these seeming benefits can become liabilities. Stalkers and assassins can use such systems to track their prey.

Beyond the daily calculations of cost and benefits, convenience and entrapment, lay the larger questions of where appropriate limits to personal privacy end and the need for public knowledge and accountability begins. In terms of knowledge, it appears that the public is little aware of the capabilities that mobile communication systems have regarding the tracking of movement and calls (Privacy Clearinghouse, 2009). And even among the technologically savvy ones there appears to be very little concern about privacy loss. In fact, even as the commercial surveillance applications grow in comprehensiveness, the general public complains but little, in contrast to the professional advocates of privacy rights, who do make their concerns quite clear. In fact, the public generally evinces high enthusiasm for trading off informational privacy for small financial benefits, such as affinity cards at supermarkets. Moreover, there seems unlimited enthusiasm for surveillance as the public finds comfort and security in omnipresent surveillance.

The Federal Communications Commission's (FCC) "enhanced 911" policy mandates that cell phone providers be able to pinpoint the location of 911 calls to expedite emergency response time. Since cell phones are essentially miniature radio transmitters and receivers, providers either use GPS devices embedded in the cell phone or triangulate coordinates from cell towers to determine a phone's location. This data has made commercial surveillance applications possible, but also enables the federal government and third parties to spy on cell phone users. The Obama administration has argued that "warrantless tracking [of cell

phones] is permitted because Americans enjoy no "reasonable expectation of privacy" in their—or at least their cell phones'—whereabouts"
(McCullagh, 2010).

This federal stance has allowed law enforcement to apply cell phone
location data to a wide variety of situations and has led to convictions
in kidnapping, drug-trafficking and robbery cases (Doyle, 2002). On
the other hand, it also allows the government to surreptitiously monitor
citizen communication. The USA Patriot Act, passed in response to the
September 11, 2001 terrorist attacks, provides the federal government
with wider latitude in electronic surveillance operations. Not only are
cell phone providers under pressure from the U.S. government to reveal customers' location information, many cell phone providers also
do not own their cell towers, which leaves customers' location data
vulnerable to the third parties that own the towers (Soghoian, 2008).
Without proper oversight, this creates a greater potential for misuse as
information could be sold for profit to other governments or spies by
the third parties.

Yet, as mentioned above, many users seem to be concerned more
about what other people in their social universe might know about them
(friends, neighbors, or personal enemies) than either Big Brother or Big
Corporation (Katz & Tassone, 1990). In fact, rather than being concerned
about protecting their privacy, many want to have others know about their
doings and their whereabouts. This is, for instance, the secret of social
networking sites. These have readily expanded to dating sites, and there
are frequent launches of new ways to share personal information, often
with strangers. There have been perennial attempts to mix mobiles with
the meeting and greeting of others, both friends and potential friends.
This adds an important positional dimension, and mobile communication
technology is of course about the only way that instantaneous locational
publicity and self-publication is possible. Mobile communication services animate and mobilize what has long been a basic human pastime:
sociality. For instance, in 2010 more than 30 percent of Smartphone users were accessing mobile social networking sites. The rapidly growing
microblogging service Twitter attracted nearly 6 million mobile users
in early 2010 (ComScore, 2010). As technology advances and services
develop, these services will likely continue to proliferate. The growth
will cast into sharp relief questions of social policy concerning how to
control behavior and what can be done to address problems of stalking
and prostitution among other topics.

M-Government and M-Voting

Mobile-enabled government, or m-government, holds great promise to solve the many ills as governmental entities seek to be more responsive to citizen needs and operate more efficiently and effectively. Despite the widespread success of the mobile phone, many governmental service applications, which would appear to have been easily implemented, have been surprisingly slow in becoming widely available. Interestingly, it seems that m-government often is more readily adopted in developing countries than in industrialized ones. In this section, we sample a few such programs and their functions.

Some programs provide information to citizens so that they can take action on their own. In Malaysia for instance, the ministry of agriculture sends Short Message Service (SMS) to farmers' mobile phones alerting them of rising water levels, thus giving them early warning to minimize damage to their fields (Ghyasi & Kushchu, 2004). Other services are designed to make life easier for citizens. In Oman, where over half the population uses cell phones, the Muscat Municipality developed an m-parking system that enables motorists to pay parking fees via SMS (Naqvi and Al-Shihi, 2009).

As discussed above in regard to privacy, mobile technology is being recruited in the fight against terrorism. In Ireland, people can use Multimedia SMS (MMS) to send photos of suspected criminals to law enforcement agencies (Ghyasi & Kushchu, 2004). Parking tickets may be paid via SMS text in countries such as Singapore, and in India one's property taxes can be forwarded via text message. In Estonia, cell phones can be used to alert city officials to broken streetlights or apply for government benefits. In a time-saving step, Estonians can also make appointments with physicians in their national health insurance plan, which also leads to more efficient use of healthcare resources on the part of staff. In one city near London, residents can send cell phone photos of graffiti defacements to local officials (Sylvers, 2008). The concept of m-government can be extended even to units at the neighborhood level, such as schools. In the United States, Korea, and Japan, parents can get text messages about their children's attendance and tardiness at school, and can receive updates about student grades as the information gets entered in the teachers' grade book.

Research studies play an important role in planning in some contexts. The results of such research can be vital for successful widespread adoption of m-government services. For instance, Use-Me.gov conducted several pilot studies across different European countries (Martín, González, González, Vilas, n.d.). In the city of Bologna, Italy, the City Information Broadcast Service offered the latest up-to-date information to the citizens, namely short-term announcements concerning public road workings, road closures and other occurrences with impact on local traffic conditions and other information such as public events and, most important in Italy, notification of upcoming and current strikes. In the Portuguese city of Villa Nova da Cerveira, a Complaint Service pilot study was conducted, aimed to foster the participation of citizens in local community matters. This service encouraged people to exercise their citizenship by allowing them to report problems and complaints, as well as make suggestions from anywhere at any time. Moreover, the service was directing the complaints and suggestions to the correct entity that was responsible for the kind of problem reported (Martín, González, González, Vilas, n.d.).

In 2002, in the Philippines, the National Police introduced a text messaging system that enabled citizens to report criminal offences (Ghyasi & Kushchu, 2004). The purpose of this service was to allow more transparency into the government offices as well as to provide better emergency assistance. All complaints were sent via text messages and routed by the mobile operator to the Complaints Referral Action Center (CRAC), where they were first recorded and then dispatched to the responsible authority to take action. There were several benefits to this service. Firstly, the complaint reporting system was improved by reducing the amount of malicious or unintentional police wrongdoings, since all messages were transferred and recorded without any human intervention. Secondly, this service also increased the capacity to handle complaints by reducing the traffic and the resulting clogging of the essential police phone lines. Since its introduction in 2002, the system received around 29,000 complaints of one third were for assistance, one-quarter to report of drug or gambling activities and 13 percent to denouncing corrupt police officers. The introduction of this service enhanced the police complaints handling procedure. However, some challenges surfaced when anonymous users sent complaints, as 90 percent of the mobile phone users use prepaid phones so it was impossible to determine the reliability in light of the huge volume of complaints. Without proper justification, and without being able to authenticate the report, it is problematical (and tedious) to pursue each complaint (Ghyasi & Kushchu, 2004).

There is perennial hope that mobile phones can increase citizen participation in local governmental activities. In particular, mobile phone-based voting is something that has already taken place in the UK, Switzerland, and Korea, among many other places, and is likely to continue to spread (Rossel, Finger & Misuraca, 2006). Mobile voting seems particularly relevant to party-based elections for nominees and perhaps for other kinds of polling. Problems of security and the secrecy of the ballot have not been fully resolved, if indeed they are resolvable. The legitimacy of the electoral outcomes is quite important, and the problems of mobile voting can be critical in a close election (such as that of the 2000 presidential campaign or the 2008 Minnesota senatorial contest in the United States). Consequently, mobile voting may be unrealizable in the near term in countries such as the United States.

On the other hand, there is no question that mobile campaigning will continue to grow. During the 2008 presidential campaign, candidate Barack Obama announced his vice presidential running mate via text message. Hungary has been a notable leader in using text messaging to organize political rallies and generate support or opposition for candidates.

High rates of mobile penetration in many developing countries, along with a lack of competing installed infrastructure, generally makes m-government services a more attractive option than e-government. Yet, there are still special difficulties in the context of developing countries including illiteracy and local language barriers. Other problems are likely to include lack of skilled IT staff, miniscule budgets, ignorance of technological best practices, and, perhaps most significantly, insufficient government and policy support.

Emergency Notification in Disaster Situations

Only recently have Information Communication Technologies (ICTs) been heavily used in disaster notification and relief. In terms of disaster relief, they have become pivotal in analyzing what steps need to be taken on the ground in response to a disaster, including rescue and medical aid. As well, for those far-distant from the disaster scene, ICTs facilitate donations from concerned people and organizing the movement of relief supplies.

Recent catastrophes, such as the 2010 Haitian earthquake, have highlighted these trends: mobile communication saved many people's lives and also served as a conduit for charitable donations. SMS can also be used in crisis events to keep citizens informed and to restore the calm within the affected population. During the SARS outbreak, the Hong Kong government sent SMS to some 6 million mobile phone users to keep them calm and reduce the tension and fear from the public's hearts (Ghyasi, Kushchu, 2004).

The mobile phone can also play a unique role in alerting people as a warning system when a catastrophe unfolds. The late 2004 Asian tsunami—certainly among the worst natural disasters in world history—highlighted this possibility albeit unfortunately due to the absence of mobile warning systems in the region. Despite the 2004 disaster, there are still few warning systems that successfully communicate to populations most in need (as discussed in other chapters in this volume). Yet the gap between promise and realization notwithstanding, text-enabled mobile phones present an excellent solution waiting to be widely implemented.

Leysia Palen and Sophia B. Liu (2007) from the connectivIT Lab & Natural Hazards Center at University of Colorado have identified three information pathways for ICT design development. These are (1) communication among the population affected by a crisis, (2) communication between those in the immediate area affected by the crisis and those outside of it, and (3) communication between the public information officers and the public.

In terms of the first pathway, of paramount importance is citizen warning. When natural disasters such as tsunamis, earthquakes, floods and wildfires occur, a critical pathway is needed to disperse promptly usable information to the affected population. To maximize their utility, such systems rely not only on sensors (seismological, ocean-floor, tidal, and satellite technologies) but also accessible and affordable communication systems that operate on the local level and can be disseminated to people who will be directly affected. This final step is the so-called "last mile" problem. The last mile is defined as the last leg of delivering communication from the provider to the customer (note that the last mile is usually more than a mile, especially in rural areas). Had last-mile dissemination been available, used properly, and had adequate information deployment systems, it could have been useful in preparing for recent disasters such as the 2004 Asian tsunami.

Samarajiva et al. (2005) define *warning* as the communication of information about a hazard or threat to a population at risk, allowing them to take appropriate actions to mitigate any harm to themselves, those in their care or their property. Current research in last-mile warning systems focuses on mobile phones as ideal communication tools due to their high penetration and adoption rates even within economically disadvantaged countries. Specifically, SMS received through mobile phones have an advantage over voice calls in that one message can be more easily sent to a group simultaneously and has much lower bandwidth requirements than voice. SMS (and mobile voice) often works when terrestrial phone lines are inoperable or congested.

Countries and regions often faced with natural disasters have begun to implement certain systems that use mobile phones to warn citizens. In Japan, the Area Mail disaster information service built by NTT DoCoMo uses a communication method called Cell Broadcast Service (CBS), to distribute earthquake warnings through SMS (Teshirogi, Sawamoto, Segawa, Sugino, 2009, p. 2). Area Mail differs from packet communication and is made based on specifications by which overcrowding of the communication lines would be unlikely even during a disaster situation.

In Malaysia, a project undertaken by Universiti Teknologi focuses on monitoring water level remotely using wireless sensor network. The project also utilizes Global System for Mobile communication (GSM) and SMS to relay data from sensors to computers or to directly alert the respective victims through their mobile phones (Abdul Aziz, Hamizan, Haron, Mehat, 2008).

While much research is dedicated to improving communication once a threat is identified, studies are also looking into technology that could increase the speed and accuracy of disaster identification. The Broadband Mobile Communications Research Lab (BMCRL) (ng) at the Asian Institute of Technology in Bangkok, Thailand works with seismic sensors that transmit automated warnings to individuals in disaster areas via wireless or wire line transmission, or by the use of visual/sound alarms.

There is a continuing need for communication within the affected public during the post-impact phase. This creates a demand for yet more rescue-related information. After a disaster strikes, typically the communication infrastructure surrounding the affected site is also destroyed. At this point the main function of emergency communication is to set up networks so that survivors can establish contact with, and give their location to, rescuers. Thailand's BMCRL analysis of emergency communication identified mobile phones as the preferred device because of their intrinsic nature to be carried on a person at all times. They see the need for a system to allow regular mobile phones to become linked to intermediary satellite systems, which are of course invulnerable to terrestrial disaster. From their viewpoint, an ideal but expensive system would include high-power emergency channels available in the satellite, add-on extendable antennas for mobile phones, and hardware/software reconfiguration in the mobile units (Asian Institute of Technology). Of course this is quite an expensive undertaking, so the Institute suggests a less ideal alternative approach in which a network of a small subset of local users, e.g., police and rescue officers, carry mobile phones that are satellite capable while other mobile phones are modified so that they can communicate with these satellite capable phones. In this way, there could be formed an ad-hoc network if the disaster destroys the cellular telephone base stations.

The second information pathway, between members of the public who are affected by the crisis and those outside it, has been the focus of ICT improvements in the past. The main objective here is to develop routing capacities and algorithms to sustain service during periods of overcrowded networks (Palen and Liu, 2007).

The third information pathway, which runs between the official public information officer function and members of the public, will experience great changes with the emergence and continued penetration of ICTs. The increased use of mobile devices and digital phone cameras creates a shift from one-way to two-way or even multi-way exchange of information. Increasingly, people have become accustomed to capturing events and sharing information immediately. Governments, the media, and emergency relief organizations are realizing the potential offered by citizen involvement in content generation. These accidental "citizen reporters" can provide important information to relief workers (Palen and Liu, 2007).

To date, most of the applications available for emergency relief tend to focus on providing support for relief organizations or for citizens, but not for both. Most commercial systems such as ATLAS Incident Management System (AIMS) and Vector Command Support System are aimed at organizational use, for managing and coordinating resources during an emergency (Lanfranchi & Ireson, 2009).

However, some attempts have begun, such as WeKnowIt, an experimental platform developed by the Department of Computer Science at the University of Sheffield (2009). Its purpose is to enable citizens distributed across a region to participate in the monitoring of an incident or event. This will benefit emergency relief planners by enhancing the available real time information upon which they can base their decisions and strategies, enabling them to better react to an emergency (Lanfranchi and Ireson, 2009, p. 2). Moreover, the system seeks to combine information from varied organization and community sources to create an accurate, comprehensive picture of the situation. It would also provide crucial information to citizens to address their immediate concerns such as what grounds are open and what information is available about the status of their relatives who might be involved. Hence, the system also aims to encourage dialogue between the Emergency Responders and the affected individuals, groups and communities. However, despite the lofty goals of the project, it remains a research undertaking that is far from complete.

Perhaps the difficulties to successful implementation of the systems lie less with the technologies themselves than with the perceived costs and policy obstacles. Although SMS warning systems provide opportunities, they do not come without problems. First is the cost issue. Even if such systems operate with complete perfection, an ideal no one expects to be achieved, setting these up throughout the world would be enormously expensive. The resources that would be devoted to emergency notification for disasters that have a likelihood of occurring less than once per century could not be used for addressing more pervasive and pressing issues, such as clean water, sanitation and education. Mobile phone/SMS broadcasts have also been met with resistance from carriers, some of who complain about the costs (relatively marginal) of setting it up or their liabilities if a message fails to go out (Mollman, 2009).

Another important issue is the problem of "false positives" which means that people are warned while no disaster actually occurs. Taking action when nothing happens can be extremely costly and disruptive on both the material and psychological levels. Moreover, it takes only a few such incidents to inure people to such warnings, perhaps rendering them equally vulnerable to the very disasters that the systems were designed to alert them to.

The issue of ubiquity is also challenging. The notification might not get through if the user does not keep the cell phone on at all times, which is often the case in remote cities in developing countries where cell phone chargers are not easy to come by, and connectivity to the electrical grid is uncertain at best.

The main issues surrounding successful deployment of ICTs in natural disaster environments revolve around concerns of feasibility, usefulness, and sustainability and although the usefulness of such programs and technologies and the feasibility of mobile phone technology have been proven, the involvement of policy makers is essential for effective long-term disaster preparedness and response.

Parents and Children with Mobile Phones

As the number of children with mobile phones increases so does the plethora of parental control features offered by cell phone companies.

This is certainly the case in North America. According to a June 2009 "Common Sense Media" poll 83 percent of teens have cell phones and 53 percent have had them before they were 13 (Common Sense Media Inc., 2010). Growing teen and pre-teen use has led to social policy by parental controls services, which are offered by many mobile phone and third-party companies. In addition to locating services, parents now have the ability to: set limits on the amount of time their children talk and text per month, regulate the time of day when phones are operable, and block certain numbers from calling their children's phone. Some may argue that while these parental controls are effective in keeping costs in check, they are a simpler solution than parents actually taking the time to discuss proper cell phone behavior with their children. There are several websites with templates for parent-child contracts for cell phone use. One from about.com runs the gamut from proper cell phone care to forbidding aggressive or explicit text messages to asserting the parent's power to monitor all communication sent and received by the phone (O'Donnell, 2010). There are even programs to control teens (or even adults) who are driving cars in terms of how fast they are going or what areas they travel in.

Educational Environment

Social policy and mobile communication affect not only families, but the larger environment of young people. This includes of course the educational environment. Mobile phone policy in American schools seems to vary widely, just as it does in terms of parent-child mobile phone relations. Reasons to limit mobile phone use in schools include: cheating, distractions and "cyberbullying" (eSchool News Staff, 2008). In 2005, a ban on mobile phone use was enacted in all New York City public schools. Despite protests from parents and students that this violated civil rights and compromised student safety, the ban was upheld by the New York State Court of Appeals in April 2008. "City lawyers argued that education officials had the right to make policy decisions—'the kind government officials make all the time'—about devices students may have at school" (eSchool News Staff, 2008). Yet, despite its stance against mobile phone use in schools, the New York City public school district does not deny the positive impact mobile phones have on students. On February 27, 2008 NYC Schools Chancellor Joel I. Klein launched the "Million" Motivation Campaign—a program that gives underprivileged middle school students a free mobile phone and rewards them with talk time and phone accessories for getting good grades in school (Wexler, 2008). Ironically, the students are forbidden to bring these school-sponsored phones to school.

Other schools allow students to bring their mobile phones into the building, but place restrictions on use during the school day. For example, in Georgia some schools have enacted fines on an escalating scale and "in-school suspensions" for students who use mobile phones in class (Torres, 2009). Other schools are easing restrictions on mobile phone use because more stringent bans did not work. For example, beginning in the 2009 school year, Millard North High School in Omaha, Nebraska allows students to use mobile phones during their lunch period in the cafeteria and a select other common areas (Baker, 2009).

It is clear that there is tension at the level of policy setting, which goes to the heart of who controls which environments. Parents want to have greater control over their children, including in the classroom. Yet schools both recognize the need for and importance of limitations on mobile phones in educational settings. As in so many other areas, the mobile phone introduces some nettlesome issues of developing social policies to address a burgeoning communication technology.

Mobile Technology and Distance Learning

As noted, in Western countries the use of cell phones in school generally creates a point of friction among professors and students. Still, developing nations are actually embracing mobile technology as a method to further education, especially in areas where distance learning emerges as the only option for students to gain some formal education.

Distance learning approaches have evolved differently in various parts of the world. Some countries initiated distance learning across technologies, such as email, forum discussion and downloadable lectures, simultaneously; while institutions in other countries, such as Japan, were very slow to adopt any distance learning as their educational culture focuses on in-person interaction. However, there is growing evidence that distance education methods have acquired a level of cost-effectiveness that provides teachers and students with numerous methods for overcoming the obstacles of time, place, and learning pace, while also engaging one another in direct interaction. As a result, efforts to develop distance-learning approaches in institutions internationally have grown rapidly (Baggaley, 2008).

The number of people requiring or wanting education and training at a distance is in the hundreds of millions around the world, and many if not most of them have mobile phones, a fact that stands in stark comparison to the number with Internet connections. The contrast is still greater in areas of the world far from urban centers. In fact, it is in the developing world that some of distance learning's most optimistic hopes are being realized, comparable at some level to the way Western educational institutions discarded educational TV in favor of Web-based approaches (Baggaley, 2008). Interestingly, their Asian counterparts may reject the Web in favor of locally developed mobile phone technology applications (Baggaley, 2008). Librero et al. (2007) have described the rapid growth of an Asian distance education environment in which cell phones and texting methods are being used to substitute for inaccessible Internet-based media (for example, Web material, email). This might not be the case for all of the Asian countries: in India Ericsson has partnered with IGNOU (2009) to facilitate 3G connectivity for students' laptops, and thus provide advanced e-learning services including online consulting and peer-to-peer discussions.

Moreover, the use of mobile technology for education is a concept that is not limited to a formal institutional setting. BRAC University in Dhaka has been experimenting with ways to use cellular SMS in lieu of computers to establish two-way interactive communication with the villages, most recently for interactive distance education (Islam and Doyle, 2008). Researchers found that educational subject matter varied with the location of the villages and the age of the villagers. Regardless of location and age, they distinguished four domains in which SMS technology can play a part: needs assessment, testing, course monitoring, and educational engagement.

Before mobile devices, like smartphones and wireless PDAs, are fully integrated into the learning process, there are a few barriers to overcome. The devices must be universally available and affordable to all students, and must have enough functionality to support a variety of inputs (handwriting, voice), communication and search capabilities. The teaching workflow must be designed to include these devices and human interaction between teachers and students. Beyond pedagogical considerations, there might be attitude, legal and ethical factors to consider. The concerns common to all mobile services, i.e., identity, access, security and privacy must be also addressed.

M-money and M-banking

In developed countries, mobile banking or m-banking most often means having access to one's bank account while on the move. This includes checking account balances, monitoring transactions and paying bills electronically via one's mobile device. However, in Third World countries, much of the population finds it impractical or unaffordable to participate in banking due to access, fees, identity, and locale, among other issues. In the case, m-banking applications that facilitate money transfers are revolutionizing the lives of societies and economies. In this domain, the rapidly growing use of mobile communication systems for banking and money transfers, and even loans, are raising intriguing social and regulatory policy issues, as the following examples demonstrate.

In 2007 Safaricom, a subsidiary of Vodafone, conducted its first money transfer via mobile phone in Kenya. The application, called M-Pesa, began as a pilot program funded jointly by Vodafone and the UK Department for International Development [DFID] Financial Deepening Challenge Fund (ITNewsAfrica.com, 2010). The introduction of applications like M-Pesa, which allow cell phone users to make and receive payments and withdraw money from virtual accounts is revolutionary considering that, in 2007 more than 80 percent of Kenyans were excluded from the "formal financial sector" (Rice, 2007). "We are effectively giving people ATM cards without them ever having to open a real bank account," said Michael Joseph, the chief executive of Safaricom (Rice, 2007).

Access to banking institutions in rural areas of countries like Kenya is limited, if not impossible. Many banks have account minimums that also preclude many workers from opening bank accounts. With M-Pesa, to make a money transfer all one has to do is: 1) give money to a registered M-Pesa agent (such as a retailer) which is then credited to their virtual account, 2) send a text message to the desired recipient, and 3) the recipient cashes in the transfer by going to an authorized agent and entering a special code and showing ID (Rice, 2007). Although Safaricom does charge a fee for transfers, it is less than banks charge, making the service even more appealing to consumers. By 2008, the service was also being used to pay employees so that employers would not have to take the "dangerous risk of carrying around bags of cash on payday" (Cheng, 2008). Safety has been an important force in the adoption of m-banking applications. Not only do m-banking applications render carrying around large quantities of money obsolete, they also allow people more privacy as to whom they are giving money.

In 2008, Vodafone announced its partnership with Western Union Co. as it expanded its m-banking ventures to include international money transfers (Sharma, 2008). International m-money transfers appeal to the many that have emigrated but send money to family members still living in developing countries. Since its 2007 inception, M-Pesa now has ten million users across Kenya, Tanzania, and Afghanistan (ITNewsAfrica, 2010). In February 2010, Hormuud Telecom introduced its own money transfer service in Somalia where remittances total about one billion dollars a year, keeping many Somalis alive (Reuters, 2010). Somali "Central Bank" governor Bashir Issa Ali told Reuters in 2010 that he believes m-banking applications are the "future banks of Somalia." Standard Bank in South Africa also launched an m-banking application called "Instant Money" to compete with M-Pesa. The "Instant Money" application will allow consumers to go to any Spar convenience store and withdraw money from their accounts (TechCentral, 2010). However, Vodafone remains the leader in m-banking in Africa by allowing for both international and domestic money transfers.

In addition to money transfers, Safaricom also offers a service called "Sambaza," which allows consumers the ability to send airtime to friends and family via text message (Safaricom, 2009). When launched in 2005, some believed the service had the potential to become an alternative form of currency: "a cyber currency that can be sent anywhere in the country at the press of a button, without needing a bank account or incurring high bank charges" (Day, 2005). In 2008, elections disputes led to violence in Kenya, which forced many shops to close. There are reports that prepaid phone cards became scarce and people instead started sending each other phone minutes via text message (Van Mensvoort, 2009). The minutes were not only used for talk time, they were used as currency (Van Mensvoort, 2009).

The relative ubiquity of ATM machines and banks is perhaps one of the main reasons m-money has not attained the same popularity in the U.S. However, there are a few companies such as PayPal and Obopay that have made some inroads in this sector. In fact, Obopay capitalizes on the wide availability of ATMs and bank accounts in America by allowing consumers to link their virtual Obopay accounts to a prepaid debit card for instance access to their money (Obopay Inc, 2010).

Yet contrast to the US and other industrialized countries, the economic and social policy implications of a parallel financial system, run through mobile phones, are enormous.

Mobile Communication and Healthcare

Based on casual conversation, it seems that when mobile communication and healthcare are juxtaposed, the conventional thought is often directed towards possible population health risks from exposure to electromagnetic fields. That is, people immediately think about the seemingly harmful effects of mobile phone communication on our health and how our safety is protected. This is indeed an important element in the problem space of mobile communication and health. However, what we mean here is the way in which mobile devices can be used to solve health problems, plus fit within the larger framework of social policy.

Generally speaking all m-health is a subset of e-health (See Rice and Katz, 2000). In many ways, m-health is able to add to the many of the virtues of e-health by making them mobile rather than institution or home based. Varshney (2007) proposed a future of mobile and ubiquitous pervasive healthcare that would offer "healthcare to anyone, anytime, and anywhere by removing locational, time and other restraints while increasing both the coverage and the quality." In this world, mobile communication is used, from the patient or public's viewpoint, for identifying problems, monitoring developments, notification of problems, conducting tests, resolving problems or administering medicine. They can easily be extended to more efficient patient-physician interaction and office visits. For the healthcare professional, mobiles are often seen, in addition to the above, as patient record repositories, instant access to health information, logistical support and instant communication with patients, peers or experts.

What is perhaps of greatest interest here is the way various mobile devices have been applied to solve healthcare problems. These are of interest less because of what they attempt to do, and their success or failure, but what they can tell us about the mobile communication-healthcare policy nexus.

There have been numerous attempts to use mobile devices to enhance healthcare. For instance, UK researchers have experimented in obesity treatments using a mobile phone that shared activity information among groups of friends. They found that awareness encouraged reflection on, and increased motivation for, daily activity. However, they also uncovered problems with network reliability related to such applications (Anderson, Maitland, Sherwood, Barkhuus, Chalmers, Hall, Brown, & Muller, 2007). A US-based study looked at having users self-monitor caloric balance in real time using a mobile phone. This was done as part of an attempt to modify user behavior to reduce obesity. In this study, Tasi, Lee, Raab, Norman, Sohn, Griswold, & Patrick (2007) conducted a one-month feasibility study to measure compliance and satisfaction among a sample of fifteen participants randomized to one of three groups. They concluded that the mobile phone was as good as or even superior to paper-based systems. The preliminary results suggest that the mobile phone could be helpful, but by no means is a panacea.

Another kind of mobile device, the radio frequency identification device, or RFID, is gaining much attention. The technology is being used in areas such as laboratory analysis, mammograms, blood transfusions, medication delivery, prostate treatment, and LASIK eye surgery (SkyeTek, 2006). Many hospitals and clinics have switched to integrated readers for barcodes and RFID chips to manage and move their inventory. In addition to commercial applications, RFID technology is also the subject of several research projects. An RFID project was trialed at a Taiwanese hospital in conjunction with a national effort to combat SARS. The project demonstrated the feasibility of RFID in hospitals but also highlighted not only technical difficulties but also both the difficulty of persuading medical professionals to accept and use the system (Wang, Chen, Ong, Liu, Wen, 2006).

One theme that comes through from several of these studies is that attitudes towards mobile communication-based healthcare systems, level of threat perceived by new technology, and one's position in a network of social relationships, all affect user evaluations and perceptions. A major concern for the m-health solutions are clinicians' acceptance and adoption of the new technologies. To gain acceptance, the whole solution, including network, hardware and software applications, should be fully integrated into one's workflow. The solutions need to address the clinician's ergonomic needs as well as handle the information needs at the point-of-care. For these technologies to be effectively designed there should be sufficient consultation with the end users (clinicians, nurses, patients) on how the application will impact their practice. Addressing these concerns is of course important to the acceptability of various technologies and policies.

Moving to the next level of analysis, it may be noted that data have been collected in several countries on public and professional responses to both specific mobile and e-health technologies and more generalized concerns. These studies have shown that fears about, for example, RFID devices for healthcare are not limited to the privacy concerns of patients and consumers but spill over into a wide range of issues. For instance, health care providers, such as nurses, are concerned about workplace monitoring and increased workloads (such as operation and maintenance) associated with RFID systems in hospitals and health care institutions, and these should be included in any implementation and evaluation of such systems (Fisher and Monahan, 2008).

A 2006 EU online survey (thus not a random sample) concerning RFIDs found that two-thirds of the 2,190 respondents believed that EU data protection and privacy legislation was currently inadequate and that stronger privacy protection laws were needed (ElAmin, 2006, & Wessel, 2006). A multi-nation series of focus groups conducted (set in Asia, Europe, and US) in 2003 found that there was little knowledge of RFID technology and that even once the benefits were discussed there was a resigned, negative sentiment towards them. In particular, for the United States, the analysis concluded that the "overall response was neutral to negative" (Duce, 2003, p. 8). These studies show that public knowledge or even awareness of RFID technologies is low. Yet despite (or perhaps because of) this, fears about the technology's privacy threats are high.

As previously stated, what is specific to m-health within the field of e-health, is that it is present in mobile environments; it can service geographic areas with limited connectivity and is delivered through devices that have limitations in power or interface capabilities. This mobile, ubiquitous characteristic endows these applications with special benefits. The major driving forces behind the development of m-health are improving patient safety, reducing costs, complying with clinician's workflow and reducing the time to access and update relevant information (Yu, Wu, Yu & Xiao, 2006).

Locales where mobile healthcare access would have major benefit include remote places, high-security healthcare facilities such as in prisons, post-hospital patient monitoring, home-care monitoring, patient education, and continuing education (Lin, 1999, pg 2) among others.

The adoption of m-health however is by no means limited to countries with highly developed health care systems. The delivery of healthcare in populated underdeveloped countries can take advantage of the quite extensive penetration of the mobile phone networks, and of the fact that users are versatile in using modern cellular phones. Relevant activities would be logistics (i.e., ordering of medical equipment and medication), performing remote diagnosis, and conducting health education and behavioral campaigns. Examples of more specific activities are campaigns to help HIV/AIDS patients, the ordering of malaria vaccinations, or helping local nurses with diagnosis when a doctor is not immediately present.

Mobile healthcare is particularly pertinent to Africa's 2.2 billion mobile phone users. In this context, a key problem is to find ways to extend mobile phone as a pathway for healthcare management, which is uncommon in Africa compared to its heavy use in people's personal activities and business (Morris, 2009). In an interesting twist, the enthusiasm for the social uses of the mobile combined with scarce economic resources have created an opportunity to do "viral" marketing of health services.

This occurs in South Africa, where a mobile operator provides a service known as "Please Call Me," or PCM. For instance, a young man who has exhausted his budget for airtime can send someone he knows a free PCM message asking her to call him back. The message sends his number but also a small space for more text. In this case, Project Masiluleke puts a health message about HIV and tuberculosis counseling in the remaining space. They send about a million such messages a day, and calls to their help line tripled (BonTempo, 2010).

Another aspect where mobile technology can be used to improve public health and healthcare is in detecting and responding to disease outbreaks. Several trials have been encouraging. For instance, in Pumwani, Kenya, a simple, low-cost mobile phone-based system has been developed to help patients with managing their AIDS (Lester, Sarah, 2008). Nurses send weekly short message service (SMS) text messages to clients receiving antiretroviral therapy to gather information on their status, and then provide pre-customized responses depending on patients' needs. Patients in the initial pilot trial gave favorable feedback on the system, reporting that "it feels like someone cares." The central advantage to using mobile phones to track patients is that they offer a cheap, quick, and ubiquitous medium across a wide variety of locales. The goal is to improve effectiveness and efficiency of health service delivery with the farthest possible reach (Lester, Sarah, 2008).

Similar to the uses of mobile technology to test students in the US, another aspect of m-health is the use of cell phones for health education in developing nations. Text to Change, a Dutch non-governmental organization (NGO), has created an SMS quiz on HIV/AIDS for Ugandans. The program not only offers free airtime but mobile prizes (Morris, 2009). The program has reported that it has stimulated a 40 percent increase in HIV testing (Morris 2009).

It is interesting to note the importance of the voice rather than the data communication capability for these areas with higher rates of illiteracy. At the same time, text is a predominant mode of use for mobiles in these areas, so it is likely that text messages would be even more efficacious than relatively expensive voice transmissions. Moreover, as graphical interfaces become increasingly common and capable, this modality will surely be tapped. In remote areas with no cellular access but where the needs are the same, the only alternative is the use of satellite communication. Here, m-health has driven the development of special technologies to facilitate communication, and even the development of new statistical distributions to model the communication channel.

A different example of an environment that awaits greater benefits from m-health is the hospital in developed countries. Here there is a need for various medical devices to transmit continuous readings to a central monitoring station, and for doctors to be able to access this information in real time from any location.

Additionally, doctors expect to be able to access written information as well as images from mobile devices, whether it is a medical image or a search in an information repository. They also expect to dictate their diagnostics into devices that translate to text and transfer the information to the health delivery chain, or capture the information in hand writing and then do the translation. Also collaboration for diagnosis can be facilitated by including local and remote members of the medical team, even the ones that are traveling. In today's hospitals and clinics we find (in increasing order of sophistication) extensive use of beepers, cell phones, electronic PDA and PC tablets.

However, in the practical application of such techniques some legal issues arise: whether medical treatment through a communication medium can be legally acceptable, and whether the transmission of medical data violates the protection of personal data (Shimizu, 1999). In the near future, it will be these issues that are the focus of policy makers.

A final element is public acceptability of invasive technology, such as RFID chips. These are the ultimate use of mobile devices, as they can be put inside a person to monitor internal states or find problems in the body, such as in the digestive tract. In what is perhaps the only national representative sample of the issue, at least in the United States, a 2007 national public opinion survey of 1,404 Americans revealed variations in interest in and desirability of several mobile healthcare technologies (Katz and Rice, 2009). Happily for the m-healthcare industry, despite the cries of alarms of critics concerning the possible dangers of these technologies generally, and of RFID in particular, there does not seem to be high levels of public concern about them. However, neither does there seem to be overwhelmingly strong positive interest. At the same time, the physical placement (not including actual insertion into the body) of potential RFID health care devices does not seem to arouse public concern except for a small minority (among those who are strongly negatively disposed toward such applications). In those cases, the negative sentiment often appears heightened when the mobile healthcare application is offered through placement on the body (such as being taped to the arm).

The results suggest high levels of public interest in emergency intervention services, but much less so in health information and monitoring services. Statistically, those who are physically closer to family and friends with whom one keeps in touch, have a stronger belief in basic privacy rights. These correlates of interest have been identified in the prior literature with interest in, or adoption of, other telecommunication services and technology. In sum, public and consumer resistance to new mobile healthcare technology may not be as great as implied by previous non-random surveys and focus groups.

Perhaps the largest lag is between the technological possibilities and the moral and policy orders and the professional interests they represent. Or as Irving Louis Horowitz describes it elsewhere in this volume, a contest between technological rabbits and moral turtles.

Conclusion

In this chapter, we explored a few domains affecting and affected by mobile communication on a US and international level. Our highly selective analysis addressed (1) privacy, anonymity and publicity, (2) m-government and m-voting, (3) emergency notification, 4) parents and children with mobile phones, (5) m-education, m-banking and (6) m-healthcare.

Although the technology of mobile communication is new, the issues that are highlighted in this chapter are in many ways perennial. What is different is that the tools to pursue them have altered both the processes and outcomes at the policy and behavioral levels. What is also different is that mobile technology has transformative potential in each of these domains. At the same time, we see that behavioral norms in certain areas, namely privacy, anonymity and publicity, seem to be changing rather dramatically. In contrast to the emergence of areas such as m-government, m-voting, m-education and m-healthcare, which developed out of a practical need for societal improvements, changes in privacy, anonymity and publicity occurred inevitably as a result of the prolific use of mobile technology. As noted in the chapter, most people are willing to reveal their location and travel habits for the convenience of immediate knowledge. Moreover, the ability of mobile devices to aid in criminal apprehension and the exculpation of the innocent are additional powerful motives for the collection and analysis of mobile communication data over and above the right to privacy. The growth will cast into sharp relief questions of social policy concerning how to control behavior and what can be done to address problems of stalking and prostitution among other topics. So the question of control of personal information becomes an important crosscutting issue of social policy.

In terms of disaster relief, ICTs have become pivotal in analyzing what steps need to be taken on the ground in response to a disaster, including rescue and medical aid; as well, they facilitate donations and organizing the movement of relief supplies. The main issues surrounding successful deployment of ICTs in natural disaster environments revolve around concerns of feasibility, usefulness, and sustainability. Most importantly, the involvement of policy makers is essential for effective long-term disaster preparedness and response. That said, a perennial issue is to what degree people need to be warned about impending danger. The situation is made more complex by the intersection of two often-unrelated sets of attitudes. The first is to what extent people want to know about the "false positives," that is the probability that something will happen as opposed to the absolute certainty that it will. Obviously, too many "false positives" will waste resources and take its psychological toll. The second is that, due to mobile warning systems, there will be a realization that these dangers are present. This in turn could increase pressure for governmental policy interventions to prevent or ameliorate disasters.

Mobile communication also affects the educational setting. Policy development in this field has already begun, as exemplified by the New York State Court of Appeals, which upheld a ban on cell phones in public schools despite protests from parents and students. On the other hand, developing nations are embracing mobile technology as a method to further education, with particular emphasis on distance learning, as an option for remote students to gain some formal education. (Of course in many developing countries there are severe restrictions on cell phone use in classrooms, or even on school grounds. On the other hand, is also the case that many examples of the most enthusiastic adoption of mobiles come from the developing world.) Hence, at the policy level in countries such as the United States, sharp challenges are being leveled at the educational institutions in terms of their imperviousness to outside influence and communication systems. Before the mobile phone and Internet, most children have had almost no access to outside communication during the school day. In the era of mobile communication, that has changed: schoolchildren persistently send and receive text messages while holding the device under their desks, updating the hoary message-passing activities from days of yore. Teens also commonly record classroom antics with their video capable mobile phones, including some that get their teachers in great trouble. For their part, parents want greater intervention in the school system that is now made possible by mobile-enabled access to their children's records, movements and activities. Thus the school has become a much more porous institution, dramatically increasing the accountability of parents, teachers, and students.

There have been numerous attempts to use mobile devices to improve healthcare both in highly developed medical settings and across the poorest areas of developing nations. Regardless of the technological challenges and social adoption concerns, issues addressing the use of mobile devices and how this influences patient privacy and physician liability surfaces as a main point needing future policy development. This would be one of the areas where there is the greatest gap between what the mobile could do for patient care and healthcare improvement and where, at least in the United States, rules of privacy, accessibility and liability prevent physicians from allowing their patients to make full use of them.

Despite some individual challenges faced in each of these areas, which we have described in this chapter, it would appear that the common obstacles slowing the mobile initiatives are concerns around privacy and security, and the lack of adequate legislation, government policies and programs to support them. Moreover, there is the question of institutional capacity: while business and social, and even political mobilization uses of mobiles have burgeoned, use for involvement with policy organs of government has been more difficult and uncertain. These issues, and many others, will be explored through varying lenses in the chapters that follow.

References

Anderson, I., Maitland, J., Sherwood, S., Barkhuus, L., Chalmers, M., Hall, M., Brown, B., & Muller, H. (2007). Shakra: Tracking and sharing daily activity levels with un-augmented mobile phones, Mobile Networks and Applications 12 (2-3), 185-199.

Baggaley, Jon. (2008). Where did distance education go wrong? *Distance Education. 29 (1),* 39-51.

Baker, J. (2009, August 13). *New School Cell Phone Policy*. Retrieved March 10, 2010, from WOWT.com: http://www.wowt.com/news/headlines/53165407.html.

Barnes, G, & Williamson, D. (2009, January 7). Athol woman and granddaughter found in Virginia. *Worcester Telegram and Gazette*, Retrieved from http://www.telegram.com/article/20090107/NEWS/901070289/1116.

BonTempo, James. 2010, (March 28). Cell phones, ubiquitous in the developing world, can be used to save countless lives. Baltimore Sun. Retrieved March 29, 2010 from http://articles.baltimoresun.com/2010-03-28/news/bal-op.cellphone0328_1_mobile-technology-mobile-phones-cell-phones.

Broadband Mobile Communications Research Lab (BMCRL). (n.d.). Emergency Communications during Natural Disasters: Infrastructure and Technology. *Asian Institute of Technology, Bangkok, Thailand.*

Cheng, Jacqui. (2008, December 8). *Cell phone money transfers going international.* Retrieved March 30, 2010, from Ars Technica: http://arstechnica.com/gadgets/news/2008/12/cell-phone-money-transfers-going-international.ars.

Choney, S. (2008, August 8). *Wireless services add more parental controls.* Retrieved March 1, 2010, from msnbc.com: http://www.msnbc.msn.com/id/26074860.

Common Sense Media Inc. (2010). *Cheating goes hi-tech.* Retrieved March 10, 2010 from Common Sense Media: http://www.commonsensemedia.org/cheating-goes-hi-tech.

ComScore (2010). Facebook and twitter access via mobile browser grows by triple-digits in the past year. Retrieved from http://www.comscore.com/Press_Events/Press_Releases/2010/3/Facebook_and_Twitter_Access_via_Mobile_Browser_Grows_by_Triple-Digits/%28language%29/eng-US.

Day, Peter. (2005, July 4). *Mutual Benefits of Profit from Poverty.* Retrieved March 31, 2010, from BBC News: http://news.bbc.co.uk/2/hi/business/4648049.stm.

Doyle, C. (2002, April 18). The USA patriot act: a sketch. *CRS Report for Congress,* Retrieved from http://www.fas.org/irp/crs/RS21203.pdf.

Duce, H. (2003). Public Policy: Understanding Public Opinion, Cambridge University, Cambridge, Auto-id Centre Institute for Manufacturing. Retrieved from http://www.rfidconsultation.eu/docs/ficheiros/Understanding_Public_Opinion_autoIDcentre.pdf.

ElAmin, A. (2006). EU public wary of RFID, privacy survey finds, Food Production Daily. Retrieved October 23, 2006 from http://www.foodproductiondaily.com/news/ng.asp?n=71489-rfid-privacy-supply-chain.

eSchool News Staff, e. a. (2008, May 29). *Court upholds New York City's cell-phone ban in schools.* Retrieved March 10, 2010, from eSchool News: http://www.eschoolnews.com/2008/05/29/court-upholds-new-york-citys-cell-phone-ban-in-schools/.

Fisher, J. A. and T. Monahan (2008). Tracking the social dimensions of RFID systems in hospitals, International Journal of Medical Informatics 77 (3), 176-183.

Ghyasi, A. Farshid. & Kushchu, Ibrahim. (2004). m-Government: Cases of Developing Countries. *mGovLab.* Retrieved from: http://www.mgovlab.org.

Harvard Law. (2004). Who knows where you've been? Privacy concerns regarding the use of cellular phones as personal locators. *Harvard Journal of Law & Technology 18,* 1 Fall.

ICA (2007). ICT as a Social Marketing Tool: Lessons from Tsunami Relief Efforts in Asia. Submission of manuscript to: ICA 2007.

Information Policy. (2009). Ericsson and IGNOU partner for 3G mobile education services. Retrieved from: http://www.i-policy.org/2009/11/ericsson-and-ignou-partner-for-3g-mobile-education-services.html.

Islam, Yusuf Mahbubul and, Kenneth Doyle. (2008) Distance Education via SMS Technology in Rural Bangladesh. *American Behavioral Scientist 52 (1)* 87-96.

ITNewsAfrica. (2010, March 30). *Vodacom and nedbank to launch m-pesa in South Africa.* Retrieved March 30, 2010, from ITNewsAfrica.com: http://www.itnewsafrica.com/?p=6595.

Katz, James E. & Annette Tassone. (1990). Public opinion on computer and telecommunications privacy. *Public Opinion Quarterly, 40* (Spring), 125-43.

Katz, James E. & Ronald E. Rice. (2009). Public views of mobile medical devices and services: A US national survey of consumer sentiments towards RFID healthcare technology, *International Journal of Medical Informatics, 78,* 104-14.

Lanfranchi, Vita. & Ireson, Neil. (2009).User Requirements for a Collective Intelligence Emergency Response System: WeKnowIT. HCI 2009 - People and Computers XXIII – Celebrating people and technology.

Lester, Richard & Karanja, Sarah. (2008). Mobile phones: exceptional tools for HIV/AIDS, health, and crisis management. *The Lancet Infection Diseases. 8 (12)*, 738-739.

Lin, James. (1999). Applying Telecommunication Technology to Health-Care Delivery *IEEE EMB Magazine 18 (4)*. pp. 28-31.

Martín, A. Solano., González, J. Mesa., González, J.A. Mesa. & Vilas, A. Méndez. (n.d.). Use-me.gov: usability- driven open platform for mobile. *Formatex Research Centre Government.*

McCullagh, D. (2010, February 11). Feds push for tracking cell phones. *CNET News*, Retrieved from http://news.cnet.com/8301-13578_3-10451518-38.html.

Morris, Kelly. (2009). Mobile phones connecting efforts to tackle infectious disease. *The Lancet Infection Diseases. 9 (5)*. 274.

Naqvi, Syed Jafar & Al-Shihi, Hafedh. (2009). M-Government Services Initiatives in Oman. *Issues in Informing Science and Information Technology. 6.*

Obopay Inc. (2010). *Obopay-go mobile with your money*. Retrieved March 30, 2010 from Obopay: https://www.obopay.com/consumer/Welcome.do?target=WelcomePage.

O'Donnell, J. (2010). *About.com: tweens*. Retrieved March 23, 2010, from A Sample Cell Phone Contract for Parents and Tweens: http://tweenparenting.about.com/od/tweenculture/a/Parent-Child-Cell-Phone-Contracts.htm.

Palen, Leysia & Liu, Sophia B. (2007). Citizen Communications in Crisis: Anticipating a Future of ICT-Supported Public Participation. *connectivIT Lab & Natural Hazards Center University of Colorado, Boulder.*

Privacy Rights Clearinghouse. (2009, August). *Privacy in the age of the Smartphone*. Retrieved from http://www.privacyrights.org/fs/fs2b-cellprivacy.htm.

Reuters. (2010, March 4). *Mobile phone money transfers offer safety in midst of conflict*. Retrieved March 29, 2010, from InsideSomalia: http://insidesomalia. org/201003042856/News/Business/Mobile-phone-money-transfers-offer-safety-in-midst-of-conflict.html.

Rice, Ronald E. & James E. Katz. (Eds.). (2000). *Internet and health communication: Experience and expectations*. Thousand Oaks, CA: Sage.

Rice, X. (2007, March 31). Kenya sets world first with money transfers by mobile. Retrieved March 29, 2010, from The Guardian: http://www.guardian.co.uk/money/2007/mar/20/kenya.mobilephones.

Rossel, P., Finger, M. and Misuraca, G. (2006). "Mobile" e-Government Options: Between Technology-driven and Usercentric. *The Electronic Journal of e-Government. 4(2)*. 79-86.

Safaricom. (2009). *Products & Services: Sambaza!* Retrieved March 30, 2010, from Safaricom: http://www.safaricom.co.ke/index.php?id=244.

Samarajiva, R. Knight-John, M., Anderson, P. S., Zainudeen, A. (2005). National Early Warning System: Sri Lanka. A participatory concept paper for the design of an effective all-hazard public warning system. Retrieved from: http://www.lirneasia.net.

Sharma, Amol. (2008, December 8). Vodafone, western union offers transfers via cell. *Wall Street Journal*, Retrieved March 30, 2010, from EB-SCO: http://gh9wn9pv9q.search.serialssolutions.com.proxy.libraries.rutgers. edu/?genre=article&isbn=&issn=00999660&title=Wall+Street+Journal++Eastern+ Edition&volume=252&issue=135&date=20081208&atitle=Vodafone%2c+Western +Union+Offer+Transfers+Via+Cell.&aulast=Sharma%2c+Amol&spage=B1&sid= EBSCO:Business+Source+Premier&pid=.

Shimizu, Koichi. (1999). Telemedicine by Mobile Communication. *IEEE EMB Mag. 18 (4)*. pp. 32-44.

Siegler, MG. (2009, April 15). When Google latitude stalking isn't such a bad thing. *TechCrunch*, Retrieved from http://techcrunch.com/2009/04/15/when-google-latitude-stalking-isnt-such-a-bad-thing.

SkyeTek Press Release. (2006). Patient safety increasingly reliant on RFID; Patient safety one of many applications RFID-enabled by SkyeTek's Advanced Universal Reader Architecture (AURA). Retrieved April 12, 2008 from Factiva Database.

Soghoian, C. (2008, September 8). Exclusive: widespread cell phone location snooping by NSA? *CNET News*, Retrieved from http://news.cnet.com/8301-13739_3-10030134-46.html.

Sylvers, Eric. (2008, November 21). With text-messaging, government goes mobile. *New York Times*. Retrieved from: http://www.nytimes.com/2008/12/21/technology/21iht-wireless22.1.18841479.html?_r=1.

TechCentral. (2010, March 25). *Standard bank seeks to head off m-pesa threat.* Retrieved March 30, 2010, from TechCentral: http://www.techcentral.co.za/standard-bank-seeks-to-head-off-m-pesa-threat/13610/.

Torres, K. (2009, August 10). *Schools get tougher on student cellphone use.* Retrieved March 10, 2010, from Atlanta Journal-Constitution: http://www.ajc.com/news/atlanta/schools-get-tougher-on-112478.html.

Tsai, C. C., Lee, G., Raab, F., Norman, G. J., Sohn, T., Griswold, W. G. & Patrick, K. (2007). Usability and feasibility of PmEB: A mobile phone application for monitoring real time caloric balance, Mobile Networks and Applications 12 (2-3), 173-184.

Van Mensvoort, Koert. (2009). *Essay: Virtual money – cows, coins, credit, airtime.* Retrieved March 31, 2010, from Next Nature: http://www.nextnature.net/2009/10/virtual-money-%E2%80%93-cows-coins-credit-airtime/.

Varshney, U. (2007). Pervasive healthcare and wireless health monitoring. *Mobile Networks and Applications 12,* (2-3), 113-127.

Verizon Wireless. (2010). *Family locator.* Retrieved from http://products.verizonwireless.com/index.aspx?id=fnd_familylocator_features.

Wang, S-W., Chen, W-H., Ong, C-S., Liu, L., & Wen, Y. (2006). RFID Application in Hospitals: A Case study on a demonstration RFID project in a Taiwan Hospital, Proceedings of the 39th Hawaii International conference on system sciences Retrieved from http://ieeexplore.ieee.org/xpl/RecentCon.jsp?punumber=10548.

Wessel, R. (2006). EU RFID survey shows privacy protection a prime concern, RFID Journal. Retrieved October 16, 2006 from http://www.rfidjournal.com/article/articleview/2736/1/1/.

Wexler, D. C. (2008, February 27). *News and Speeches: Chancellor Klein Launches "Million" Motivation Campaign.* Retrieved March 10, 2010, from NYC Department of Education: http://schools.nyc.gov/Offices/mediarelations/NewsandSpeeches/2007-2008/20080227_million.htm.

Yu, P., Wu, M. X., Yu, H., Xiao, G. C., (2006). The Challenges for the Adoption of m-Health. *IEEE. 18(9).* 181-186.

3

Mobile Communication
and Socio-Technical Change

John Leslie King

This chapter takes a socio-technical perspective on mobile communication. Socio-technical, to quote the ephemeral Wikipedia, means
"the interaction between society's complex infrastructures and human
behavior" ("Sociotechnical Systems," 2010). The chapter is guided by
what Bowker calls "infrastructural inversion," an attempt to overcome
the fact that infrastructure tends to disappear from view by examining closely technologies and arrangements that make infrastructure
what it is (Bowker, 1994).[1] Infrastructure constitutes an intermediating
construct between mobile communication and social change, useful
in examining change over time. The chapter does not claim that the
resulting explanations are the only way things might have turned out.
The chapter is merely one point-of-view grounded in the belief that
no single factor explains socio-technical change. The value in taking
a broader, ecological point-of-view is that it can include individuals,
organizations, technologies and policies while not letting any single
factor dominate.

The chapter tells a story in three acts. The first concerns the rise of
systems, networks and science in large-scale socio-technical endeavor
during the late nineteenth and early twentieth centuries when universal
service and the public-switched telephone network were created, scientific approaches to management blossomed, and government-regulated
enterprise emerged. The second concerns consolidation, rationalization and the paradigm of the regulated monopoly for utility provision
that took place in the middle decades of the twentieth century, during
which the AT&T Bell System dominated telephony in the US. The third
concerns market liberalization and the fundamental transformation of

communication infrastructure and use in the latter decades of the twentieth century, during which the concepts of telephony and interpersonal communication were turned upside down. Changes in technology, organizations, and institutions over this period were tied to changing public views of the relationship the individual to society and the role of communication infrastructure.

Systems, Networks, and Science

The late nineteenth and early twentieth century saw the rise of systems, networks and science in human enterprise built on deep roots in changing technology. By the late nineteenth century steamship and steam train companies were offering scheduled service via transportation networks that allowed companies to control their operations end-to-end, predicting with accuracy when given ships or trains would reach their destinations. *System* was essential in this transformation, built on new communication technologies and new ways of organizing enterprise (Yates, 1989). Information flows within companies became networked as never before, and in industries such as railroads, networks became the backbones of enterprise.

Systems and networks are seen in the relationship between the railroad and the telegraph. Telegraphy, invented by Samuel Morse in the 1830s, had become essential in the operation of the railroad by the mid-1800s. The telegraph separated communication from transportation, sundering the ancient dependency of communication on transport; a factor that influenced all subsequent electrical and electronic communication (Carey 1989). Telegraphy and railroads co-evolved. Telegraph companies found ways to inter-operate and pass messages among networks as demand for end-to-end communication grew across the entire range of telegraphy's reach (Nonnemacher, 2005). In fact, problems with early network formation eventually led to network unification and the creation of the de facto Western Union monopoly. Similarly, local railroads found ways to inter-operate with other railroads and thereby covered the map with service. Travel booking and freight forwarding had long been coordinating transport across multiple transport lines, but telegraph and railroad networks made this easier. Passengers might have to make their way from one railroad company's station to another, but telegraphy helped travel agents arrange intermediate passage and minimize inconvenience. Freight might have to be hauled between depots via drayage, but telegraphy helped forwarders coordinate that. Eventually, cities developed "union" stations and inter-connected depots so passengers and freight could "stay on steel" for the entire route. Networked systems facilitated railroad consolidation between 1860 and 1890 (Grodinsky, 1930).

Similar advances in coordination were at the heart of a revolution in manufacturing. Interchangeable parts for multi-part products were used in the manufacture of firearms by the beginning of the nineteenth century, laying the groundwork for mass production. Ransom Olds of Lansing Michigan applied the concept of the assembly line to make his popular 1901 "Curved Dash" Oldsmobile. Henry Ford adopted that concept and expanded it to production engineering in which the entire manufacture and assembly process was pre-planned and choreographed. Production engineering transformed manufacturing, dramatically lowering the price of finished goods and accelerating the mechanization of agriculture. These drove the transformation from the agrarian era to the industrial era, including a dramatic migration of the population from rural areas into cities.

Production engineering and similar innovations were not merely the result of ingenuity; they required the application of systematic knowledge gained from scientific thinking and practice. The late nineteenth century saw the rise of modern "epistemic infrastructure:" great libraries, archives, museums, botanical gardens, zoos, aquaria, and other systematic collections that fostered new knowledge communities (Hedstrom & King, 2006). In large part because of Prussia's emphasis on science, by the mid 1870s it had surpassed Britain in industrial invention and, partly by use of organizational science, defeated France in the Franco-Prussian War. Governments around the world began promoting science in ways that eventually evolved into national systems of innovation (Freeman & Soete, 1997). In the US, the Smithsonian Institution was established in 1846, and the Morrill Act creating the Land Grant institutions of higher education in the states was signed into law in 1862 ("Morrill Land-Grand Colleges Act," 2010). Universities such as Johns Hopkins and the University of Michigan embraced scientific research as essential to their missions, creating partnerships with industry and commerce in the process. Still, applied science was largely a private sector affair for many decades. Thomas Edison established the first laboratory devoted to industrial research in 1876, and by the early 1900s it was common for companies to have their own research centers (Israel, 1992). The first professional engineering association, the ASME (American Society of Mechanical Engineers) was founded in 1880.

Scientific thinking spread to the organization and administration of enterprise, broadly. The first formal school of business administration, the Wharton School at the University of Pennsylvania, was established in 1881. The concept of "scientific management" grew dramatically between 1880 and 1920. In the early 1880s, while working as an engineer in a Pennsylvania steel factory, mechanical engineer Frederick W. Taylor began working to improve industrial efficiency through careful time-and-motion studies of each step in production. He conducted such work for 20 years, served as President of the ASME in 1906-7, and published his philosophy of enforced standardization, enforced adoption of best implements and working conditions, and enforced cooperation among workers was published in 1911 as *The Principles of Scientific Management* (Taylor, 1911). The term "scientific management" was coined by future Supreme Court justice Louis Brandeis in his arguments for the Interstate Commerce Commission's Eastern Rate Case in 1910. Brandeis argued that Taylor's principles would allow railroads to increase efficiency, raise worker wages, and improve profits without raising regulated railroad tariffs.

Scientific management spread quickly. Frank Gilbreth brought principles of scientific management to the construction industry, reducing the number of steps involved in brick-laying from 18 to five. In 1904 he married Lillian Moller, credited with bringing psychology to the study of industry. The two of them worked on efficiency studies, and after Frank's death Lillian helped to establish the field of home economics (Graham, 1999). Physician Luther Gulick promoted physical fitness and the emerging sport of basketball (E.g., Gulick, 1907). In 1910 he and his wife Charlotte founded the Camp Fire Girls, the first national organization advocating physical fitness and outdoor activity for girls. Gulick later distilled the principles of business leadership into the acronym POSDCORB (planning, organizing, staffing, directing, coordinating, reporting and budgeting) and joined forces with management scholar Lyndall Urwick to create the academic journal *Administrative Science Quarterly*.

Australian Elton Mayo moved to the United States and the Wharton School in 1923 to study the effects of fatigue on worker turnover, and two years later joined Fritz Roethlisberger and his group at the Harvard Business School. They studied the effects of illumination on worker productivity in the Relay Assembly Test Room of the Western Electric Hawthorne Works outside Chicago, research that became known as the Hawthorne Studies. Neither Max Weber's concept of bureaucracy nor

Taylor's enforced standards, methods and cooperation were sufficient for them (Weber, 1978). Work took place in "informal" organizations embodying actions, norms, conventions and connections among individuals and groups that were developed over time (Harvard College, 2010). While they shared with scientific management an appreciation of close and systematic observation of actual work situations, they focused on the *context of work* as well as the actions of workers performing work-related tasks.[2]

This time also brought startling developments in the natural sciences that captured the public imagination. Charles Darwin's controversial theory of natural selection was still eliciting protests. Theoretical physicists attempted to reconcile aspects of classical mechanics using Euclidian space in a problem now known as space-time ("Spacetime," 2010). Albert Einstein provided a startling breakthrough in 1905 by suggesting that time is not a constant, as had always been thought, but was a variable: time itself is relative. Ernest Rutherford split the atom in 1917. Niels Bohr developed quantum theory in the early 1920s and Werner Heisenberg published his famous "uncertainty principle" in 1927, which had far-reaching intellectual influence beyond the world of physics. Soon thereafter Edwin Hubble demonstrated that the universe is not stationary, but expanding (Osterbrock, Gwinn, & Brashear, 1993). Science in all forms was gaining respect and public notice.[3]

Scientific rationality played a vital role in shaping industry in the late nineteenth and early twentieth centuries. For example, fixed schedules required standardized time, including the creation of time zones. This began in Britain at the Greenwich Observatory in the mid nineteenth century and eventually spread around the world. Standard time was adopted by US railroads in 1883 and Congress established time zones in 1918. Time zones allowed the alignment of routes of multiple carriers in sequence across great geographic distances. The telegraph facilitated pre-processing (e.g., booking in advance on a sequence of carriers) and exception-handling at run-time within the network (e.g., recovery from unexpected events, rebooking in the event of delay). Improved communication and information reporting systems also brought the ability to collect information at various points within the physical network, synthesize the data into standardized formats, and construct abstract models of the actual networks that could be analyzed by specialists to improve operation. These improvements were established by the railroads through use of telegraphy, and paved the way telephony under the leadership of Theodore Vail.

Vail's story includes both telegraphy and the railroads. He was the son of Alfred Vail, Samuel Morse's closest technical collaborator in the creation of modern telegraphy. Theodore was himself a telegraph operator and the superintendent of the Railway Mail Services before being appointed as general manager of the Bell Telephone Company in 1878 (Brooks, 1975). The Bell Telephone Company maintained a patent-based monopoly on telephony at a time when citizens and policy-makers were becoming concerned about monopolists. The railroad industry had formed *de-facto* monopolies along the routes and in the regions controlled by single carriers, and the "railroad barons" were increasingly the object of criticism. John D. Rockefeller's Standard Oil upset the traditional vertical marketing networks of jobbers and other intermediaries by taking kerosene directly to customers through its own tank cars and distribution wagons. This dramatically lowered the price of kerosene, which consumers liked, but the company's use of predatory pricing and other mechanisms to attack its competitors was well known. A major scandal erupted when it was revealed that the company had made secret deals with consolidated railroads for rebates on transportation of the company's tank cars through a front organization called the South Improvement Company. Such scandals sparked the regulation of commerce (e.g., the Interstate Commerce Act of 1877), the regulation of trusts (e.g., the Sherman Anti-Trust Act of 1890), and the beginning of regulation in capital markets.

Railroads, telegraphy, petroleum, electric power and other industries started as loose conglomerates of stand-alone, regional companies. Over time, connection and inter-connection among the companies' network infrastructures created broad coverage and services at lower prices than ever before. The application of science to enterprise enabled and emphasized the power of consolidation to reduce costs and improve quality and service. Consolidation also precipitated widespread public concern about control in the hands of a few powerful actors. Consolidation was a key to network success, but as consolidated network infrastructures emerged in the late nineteenth century, these industries were accompanied by growing political reaction against the large and powerful players such as Standard Oil. The regulatory reforms of the Theodore Roosevelt and William Howard Taft administrations between 1901 and 1913 were largely in response to the tremendous power of monopolies that could exploit networks, and ushered in an era of regulation that was to prevail for six decades.[4]

Consolidation, Rationalization, and Regulated Monopoly

Theodore Vail retired from his first term as Bell CEO in 1889, spent fifteen years making a fortune in mines and manufacturing in Argentina, and then re-joined the company as CEO in 1907. During his absence the original Bell patents had expired, and competition was growing. The company had acquired Western Union and had been renamed the American Telephone and Telegraph Company, AT&T. Vail's second term of leadership set a new course for telephony. Vail had an early appreciation of the power of science, and used scientific research in the creation of a long-distance network that would link together local operating companies. Sending voice signals across many miles of wire with erratic resistance and capacitance characteristics was a serious technical challenge. Vail called on the research department of AT&T's Western Electric for help, and in 1914 Vail himself placed the world's first transcontinental phone call from New York to San Francisco. The next year telephone service to Europe via transoceanic cable began. Vail's research department evolved to become Bell Laboratories in 1925.

By the early twentieth century it was clear that with networks, bigger was better. Operation of large networks required scientific research that went beyond technology to include management and organization. There seemed to be little limit to the power of scientifically informed, appropriately regulated, large-scale private enterprise. In 1908 Theodore Vail began an advertising campaign "the Bell System," accompanied by the slogan, "One Policy, One System, Universal Service." This began a process that established the Bell System as an institution in American life (AT&T, 2010). (This succeeded to a remarkable degree, with the Bell System eventually acquiring the nickname "Ma Bell" that denoted both matronly care as well as parental authority.) Vail's idea struck at a simple but important problem. Competition brought multiple, separate telephone systems serving the same customers. To talk to everyone, it was necessary to subscribe to all the services that anyone might be using. Costs were high and services suffered. In time this was articulated as a violation of a principle in economics dating back to Malthus, the "natural monopoly" that grew out of natural conditions rather than actions of people (Mosca, 2006).

One Policy
One System

Universal Service

THAT the American public requires a telephone service that is universal is becoming plainer every day.

Now, while people are learning that the Bell service has a broad national scope and the flexibility to meet the ever varying needs of telephone users, they know little of how these results have been brought about. The keynote is found in the motto—"One policy, one system, universal service."

Behind this motto may be found the American Telephone and Telegraph Company—the so-called "parent" Bell Company.

* * * *

A unified policy is obtained because the American Telephone and Telegraph Company has for one of its functions that of a holding company, which federates the associated companies and makes available for all what is accomplished by each.

As an important stockholder in the associated Bell companies, it assists them in financing their extensions, and it helps insure a sound and uniform financial policy.

* * * *

A unified system is obtained because the American Telephone and Telegraph Company has for one of its functions the ownership and maintenance of the telephones used by the 4,000,000 subscribers of the associated companies.

In the development of the art, it originates, tests, improves and protects new appliances and secures economies in the purchase of supplies.

It provides a clearing - house of standardization and thus insures economy in the construction of equipment, lines and conduits, as well as in operating methods and legal work—in fact, in all the functions of the associated companies which are held in common.

* * *

Universal, comprehensive service is obtained because the American Telephone and Telegraph Company has among its other functions the construction and operation of long distance lines, which connect the systems of the associated companies into a unified and harmonious whole.

It establishes a single, instead of a divided, responsibility in inter-state connections, and a uniform system of operating and accounting; and secures a degree of efficiency in both local and long distance service that no association of independent neighboring companies could obtain.

* * *

Hence it can be seen that the American Telephone and Telegraph Company is the active agency for securing *one policy, one system,* and *universal service*—the three factors which have made the telephone service of the United States superior to that of any other country.

American Telephone & Telegraph Company

AT&T advertisement from 1908 regarding universal service. (Courtesy of AT&T, reprinted with permission.)

Vail's concept of universal service meant that only one service provider would cover a given geographic region, providing high quality of service and lower costs of operation. Vail implemented this by acquiring competitors in areas where the Bell System could take control, and abandoning to competitors areas that the Bell System could not control. AT&T was aggressive between 1908 and 1912, using the techniques of the time to take over competitors. The company was charged with refusing to allow competitors to connect to its dominant long-distance network. Complaints were growing, and Congress was deliberating nationalizing the telephone system as other countries had done. US Attorney General George Wickersham filed an anti-trust lawsuit against AT&T in 1912. In December of 1913 AT&T Vice President Nathan Kingsbury sent a letter to the Attorney General outlining AT&T's response to the anti-trust complaint ("Letter to the Attorney General," 1914). AT&T would divest itself of Western Union, would no longer acquire independent telephone companies unless the Department of Justice and the Interstate Commerce Commission agreed (a provision eliminated seven years later), and would allow independent local networks to interconnect to its long-distance network (Sterling, Bernt, & Weiss, 2006). This proposal was accepted and became known as the Kingsbury Agreement.

AT&T began implementing the Kingsbury Agreement, building the first stages of what was to become the Public Switched Telephone Network, the PSTN. However, in 1914 World War I broke out in Europe. It was the first war in which the whole world could be involved due to changes in transportation and communication. (Practical two-way radiotelegraphy emerged before 1910, and the Titanic disaster of 1912 eventually prompted ship-to-shore radio for all passenger vessels.) Although the US did not enter the war until 1917, when it did the country experienced its first national mobilization. US railroads and the telephone system were nationalized. The war reinforced the view that national mobilization could work, but problems with the nationalization of the railroads and the telephone system also convinced citizens and policymakers that enterprises such as transportation and communication could be run more effectively by private than public organizations (Brooks, 1975). In the 1920s the US model evolved as a marriage of large-scale, privately operated network infrastructure governed by regulations to curb abuses that such power might precipitate.

The Kingsbury Agreement and AT&T's actions following World War I empowered AT&T and its Bell System to become something close to a telephone monopoly in the United States. However, it was *regulated monopoly.* It was supposed to achieve the benefits of consolidation and scientific approaches to business while avoiding the abuses of unregulated monopolists. Public commissions were created as regulators. They were typically state-level or regional bodies that would monitor quality of service and keep rates as low as possible, while providing the monopoly sufficient returns to make infrastructure investments and maintain service. Regulation embodied the concept of the "common carrier," a business that transports people, goods, or services for the general public under license from a regulator that had evolved in transport and was extended to communication. The regulated telephone monopoly as it evolved throughout the 1920s became enshrined in the Telecommunications Act of 1934 that established the Federal Communications Commission at the national level to regulate both common-carrier communication and broadcasting.

As AT&T and its Bell System grew to be a powerful and technologically proficient institution, the concept of "universal service" changed. The Telecommunications Act of 1934 included a brief description of "universal service," but it did not mean what was intended by Vail in 1908. "Universal service" in the act did not mean that anyone should be able to call anyone else through a single system, but rather that everyone should have telephone service—that the service should be universal. Areas of very low population density could be excluded as too expensive to service, but customers anywhere outside those areas would receive service from the regulated local telephone companies at a flat rate as set by the public commissions, regardless of how far they were from the central office. This concept of universal service was extended in 1935 when the Rural Electrification Administration was launched to bring electric power to rural areas. In 1949 the REA was given authority to provide assistance to telephone cooperatives to extend service to areas not yet covered (D.C. Brown, 1982). The Bell System as a regulated monopoly became an institution of scientific attitude and corresponding capability. The United States became the "gold standard" for telephone service, and the Bell System provided service beyond the borders of the US to include Canada and the other countries (RxS Enterprises, 2010).

Along the way, the Bell System engaged two technologies that were to become important in mobile communication several decades later: automatic switching of telephone calls, and radiotelephony. In the late 1800s Almon Strowger, an undertaker in Kansas City, devised an electromechanical step switch capable of processing pulsed electrical signals corresponding to the digits 0-9. An array of such switches could determine a path for a circuit. Strowger patented his device in 1891, and with some members of his family created the Strowger Automatic Telephone Exchange Company. In 1892 the company opened the country's first automatically switched telephone office in Indiana. In 1896 the rotary pulse dial was developed for telephones, making it possible for the user to instruct the switches to connect the call as desired. In 1901 Strowger's company evolved into the Automatic Electric Company, a direct competitor to Western Electric in the creation of telephone equipment and a major supplier to the independent local exchange companies for many decades. The step switch evolved and was adopted by phone companies worldwide, including the Bell System in 1919. Automatic switching was not merely a technological change; it had profound implications for the operation of the infrastructure.

The local Bell System companies originally used human operators working at manual patch bays. A light would appear to signal an incoming call, the operator would plug a patch cable into the panel at the appropriate point, talk to the calling party, get the number for the called party, and connect the calling party to the called party manually. Boys were initially hired to serve as operators, but they were often restless and rude, and were eventually replaced by women who held telephone operator jobs almost exclusively for many years. AT&T's experts in scientific management determined that, if call volume grew as predicted, there would not be enough operators to place all the calls. The company had to move toward automatic switching, but it initially did not want to buy Automatic Electric equipment. The Bell System had a practice of replacing automated switched exchanges with operator-switched exchanges when acquiring competitors, but stopped this when implementation of Automatic Electric switches began in 1919 (Privateline.com, p. 6, n.d.). Western Electric continued developing automatic switching equipment to replace the Automatic Electric system, introducing the unsuccessful panel switch in 1921. AT&T finally began replacing the step switch with the crossbar switch in the early 1940s. The vast majority of Bell System phones were converted to automatic switching by the 1950s, but the last phones were not converted until 1978.

Without automation, a large portion of the population would have to become telephone operators in order to switch all of the calls. Automation was a productivity strategy, but an odd one under the circumstances. Capital-labor substitution during the industrial revolution meant less labor was required to do an equivalent amount of work.

Automatic switches reduced the number of human operators, but only by making the population at large *into* telephone operators. The calling party could no longer ask human operator to make the connection, but had to know the number of the called party and how to execute a sequence of tasks in the right order to place the call. Automation transferred the work from Bell System operators to customers. It did not prevent the population from becoming telephone operators; it enabled and necessitated their becoming telephone operators. In the process, users gained a measure of control over the infrastructure that was to prove important later.

By the end of World War II, individuals in households, workplaces, and many public places had access to a unified system controlled by a regulated monopoly. The system was millions of telephones connected to a wireline network as fixed *stations*. Users came to a station to make a call. Residential service was a telephone fixed to the wall or sitting on a table, or perhaps in a special-built nook, tethered via a wire to a wall. The wire ran from the house to the central office. A second telephone in the residence was an "extension." Telephone instruments were not owned by the customer, but were leased by the customer from the telephone company as part of the monthly service fee. "Mobile" service was accomplished through widespread deployment of wireline public telephones, either within buildings or outside in stand-alone telephone booths. With enough public telephones an individual was likely to find a wireline *station* almost anywhere he or she happened to be. Mobile service was not truly mobile—the users were mobile but within a short distance of ubiquitously available wireline stations. Business and industrial telephone customers could operate a Public Branch Exchange (PBX), a small version of a telephone company's exchange but located on a company's premises. The PBX allowed internal telephone calls or calls to outside parties through accessing an "outside line" meaning a telephone line in the PSTN. A combination of human operators and automatic switching allowed the PBX to operate as a small, privately-owned telephone company with access to the local telephone company, and through them to the long-distance network. Companies were prohibited from taking the PBX concept too far and setting up their own telephone system by regulatory restrictions that typically prohibited the crossing of a public thoroughfare with a telephone line unless that line was operated by a common carrier. Although everything was tied together by the PSTN, the calling party was increasingly the effective operator of that network.

"Telephones are ubiquitous?"

"Yes, telephones are ubiquitous"

Robert Day, the well-known cartoonist who drew this picture, was a little afraid there might be some people who wouldn't know what we meant by ubiquitous. "It's a pretty big word," he said.

"Don't worry," we told him. "We'll just put in a little reminder that the dictionary says ubiquitous means 'existing or being everywhere at the same time.'"

There's surely no better way to describe telephones! They're not only in millions of homes and offices but just about everywhere you go. In stores and at gas stations! At airports, bus depots and railroad stations! Out-of-doors!

Throughout the country, there are hundreds of thousands of these public telephones for your convenience.

So the next time something comes up when you're away from home or the office, or there's some news you'd like to share with someone, just step up to one of those nearby telephones and call.

You can save yourself a lot of running around, be a number of places in a few minutes, and get things settled while they are fresh in your mind.

Working together to bring people together . . . **Bell Telephone System**

AT&T advertisement in Life Magazine from *LIFE* Magazine, March 4, 1957 suggesting the ubiquity of wireline telephones. (Reprinted with permission.)

Radiotelegraphy was in use by 1910, and two-way radio voice communication was underway in police departments by the 1930s. The 1934 Telecommunications Act gave the FCC jurisdiction over Radio Common Carriers (RCC) that provided two-way radio for commercial services (trucks, taxis, etc.) but service was expensive and relatively rare. The 1934 Act treated the radio spectrum as a precious and limited resource, and the FCC regulated access to frequencies. The FCC also regulated broadcasting. The rise in popularity and power of radio and subsequently television made radio spectrum an important political factor in the FCC's operations. An analog wireless telephone call took considerable radio spectrum, so only the most "worthy" applications could be handled under RCC. Such services did not fall under the same scrutiny as residential telephone service. Universal service under the 1934 vision was reserved to wireline telephony, and rates for wireless communication could be set to dampen demand rather than serve a broader population.

AT&T recognized the value of being able to connect radiotelephone calls to through the PSTN, either to wireline extensions or to other radiotelephone users. This early implementation of mobile telephone service was in operation by 1946. Radiotelephone calls went out through a single, powerful broadcast station, whereupon they could be picked up within the receiving range by radio transceivers in vehicles. The weaker broadcasts from the vehicles were picked up by distributed antennae and carried via phone lines back to the mobile telephone service center. Calls were handled by human operators. The service could not handle many calls simultaneously (fewer than 25 in most cases), meaning that the market for mobile telephone services could not grow much. In response to this constraint Bell Laboratories scientists and engineers developed the concept of cellular telephony in 1947. Cellular telephony would in principle open more channels to meet growing demand, but it was out of reach technically at that time. In the 1960s the Bell System developed the more sophisticated Improved Mobile Telephone Service (IMTS) that allowed automatic switching, but did not significantly increase capacity. Demand still exceeded supply, and the price for the service was high. Analog cellular telephony in the US finally appeared in 1983 as the Bell System's Advanced Mobile Phone System (AMPS) (Fors, 2010). By that time AT&T and the Bell System were in a new era.

During its regulated monopoly heyday, the Bell Telephone System used its circuit-switched, wireline network to fulfill the 1934 Act's concept of universal service. Individuals almost anywhere except remote wilderness would be near a *station* whereby they could place or receive a telephone call. By 1960 the calling party in most areas was the operator of this net-work, using the automated switching system to phone a called party without intervention of a telephone company operator. The Wide Area Transmis-sion System (WATS), introduced in 1967, allowed a customer to pay a base charge and make outgoing long-distance calls toll-free, or to set up a toll-free long-distance number to allow calling parties to call in toll-free (the 800 service). Before long calling parties could use calling cards and other mechanisms to make local or long-distance toll calls from any location and charge them to another number or to a credit card without using a human operator. The user was taking greater control of the infrastructure, and the user's relationship to the telephone company was increasingly abstract and limited to establishment and maintenance of service contracts. Human-to-human communication between customers and the telephone company were increasingly rare, and "Ma Bell" grew increasingly distant even as the level and array of telephone services increased.

Market Liberalization, Digital Technologies, and the Transformation of Communications

The regulated monopoly model reached a peak shortly after World War II. Mobilization for World War II for all practical purposes nationalized US industry, but soon after the war the Supreme Court signaled a return to pre-war political conditions when it thwarted President Truman's attempt to nationalize the steel industry in April of 1952. This was during the Korean conflict, and Truman argued that nationalization was necessary for national security, but his argument did not hold. The Cold War prompted many sophisticated programs under the mantle of national security, and technological progress was at the center. The Soviet Union launched Sputnik in 1957, and the US responded with Explorer 1 in 1958, and the Department of Defense created the Advanced Research Projects Agency, ARPA, in 1958 with the mission to "prevent technological surprise" ("DARPA," 2010). This was the foundation on which President John F. Kennedy based his astonishing goal of landing a man on the moon and returning him safely to Earth by the end of the 1960s in a speech to a joint session of Congress on May 25, 1961 (Garber, 2010). The decision to go to the moon was "a major national commitment of scientific and technical manpower, material and facilities" (Kennedy, 1961). The space race proceeded, and brought advances across a wide array of technical fields, some of which proved important to the development of contemporary mobile communication.

Kennedy's prophetic speech was preceded by less than four months by a different and equally prophetic speech. On January 17, 1961, Dwight Eisenhower gave his farewell address as president. Eisenhower pointed the "grave implications" of a permanent arms industry and a large standing military, both of which were new to the US experience. He warned of the "military-industrial complex" and research that was "centralized, complex, and costly." Political leaders, he said, must prevent the country's becoming captive to a "scientific-technological elite." Eisenhower's words foreshadowed an era of dramatic political change. Within four years Barry Goldwater lost by a landslide to President Lyndon Johnson, but his campaign sparked a movement that subsequently found voice in Ronald Reagan and other conservatives who favored strong defense but criticized government as "the problem." The rise of the conservative movement called into question long-standing collaborations between government and industry, and especially the regulated monopoly, and precipitated a dramatic period of deregulation.

The takedown of the regulated monopoly began in the Carter administration with the Airline Deregulation Act of 1978, passed by a Democratic Congress with a filibuster-proof majority in the Senate. That act reduced the Civil Aeronautics Board's authority to operate as a public commission regulating tariffs and routes ("Airline Deregulation Act," 2010).[5] Deregulation continued into the 1990s with the Clinton administration's agreement to eliminate the Interstate Commerce Commission. The political movement for broad-scale liberalization of regulated markets was bipartisan. It took place under the Carter, Reagan, Ford, GHW Bush, and Clinton administrations, during periods where Congress was under Democratic and Republican rule. This movement fundamentally challenged the idea of regulated monopoly, and telephony was affected dramatically.

AT&T's monopoly had been challenged over the years, eroding the company's authority but failing to break the monopoly itself. By the 1970s circumstances had changed. Microwave Communications Incorporated, better known as MCI, sued AT&T in March of 1974. MCI carried long-distance telephone traffic via microwave, bypassing AT&T's Long Lines service, and resisted AT&T efforts to stop such services as a violation of the 1934 Act. The US Department of Justice in the Ford administration followed with its own anti-trust suit against AT&T in November of 1974. AT&T protested that the 1934 Act protected it against anti-trust action, but the federal court handling the case rejected that claim. The Carter Justice Department continued the anti-trust case against AT&T, as did the Reagan administration. In 1982 AT&T engaged in an odd replay of the Kingsbury Agreement, offering this time to divest itself of the Bell Operating Companies that had given the company its start. AT&T negotiated a deal with the federal court that became the Modified Final Judgment, entered in August of 1982 (M. Brown, 1981).[6] The Modified Final Judgment took effect January 1 of 1984. AT&T and long distance telephone service were removed from regulation, and the 22 Bell Operating Companies were spun off into seven regional companies

(RBOCs) providing regulated local telephone service. The Bell System's long practice of subsidizing local telephone service through high charges for long-distance service ended, and the long-distance market became competitive (Cybertelecom, 2010). The Telecommunications Act of 1996 amended the 1934 Act, to allow competition in both local service and long distance. The seven RBOCs consolidated into two, and one of them, SBC, bought AT&T and its long distance service. This reunited local and long distance service.

Regulated monopoly had accomplished much during the twentieth century: the US led the world in telephony, motor transport and air transport, regulated respectively by the FCC, the Interstate Commerce Commission, and the Civil Aeronautics Board. The Cold War was built on the "can-do" momentum of World War II through spectacular governmental and quasi-governmental efforts such as the Manhattan Project, the deployment of the US nuclear deterrent, and the Race to the Moon that culminated in the 1969 landing of Apollo 11. Massive investment was made in research by large private companies including Bell Laboratories and IBM's many research centers, as well as government support of Federally-Funded Research and Development Centers (FFRDCs), mission agency research (e.g., DARPA), and by the National Institutes of Health and the National Science Foundation that funded university-based research. These led to technological innovations on which contemporary mobile communication was built, especially microelectronics and software.

The late 1960s also called into question many deeply-held social and political views: the sexual revolution; the civil rights movement; the war in Vietnam; the counter-culture movement; the environmental movement; the anti-nuclear movement; the women's movement. These changes did not bring about the revolution in mobile communication, but they prepared the way for changes in attitudes about the use of infrastructure. By this time powerful infrastructures had come under the control of end-users. Automobiles and the Interstate system enabled personal rapid transit. Automatic telephone switching, the advent of WATS, and the implementation of other billing-related innovations put telephone system operations in the hands of users. Home recording of music from LP disks onto compact cassettes allowed users to create programming for the Sony Walkman that appeared in 1980. (Portable CD players soon followed.) Cable television increased the number of television channels available. The Xerox 914, introduced in 1960, changed office behavior. Computers were miniaturized in the 1970s and micro-sized by the early 1980s when the IBM PC was introduced. Apple Computer introduced the graphical user interface in the Macintosh in 1984. The ARPANet came out of the obscurity of its origins in university research laboratories and grew into the Internet. *The New Yorker* cartoon of July 5, 1993 showing a dog at a computer telling another dog, "On the Internet nobody knows you're a dog," appeared two years before the name "Internet" was officially adopted.[7] It seems fitting that the Bell System was established by an intensive advertising campaign in the beginning of the century, while the Internet got its name through a cartoon at the end of the century.

Mature adults in the 1960s had grown up with regulated monopoly, and understood it as the normal order of things. They flew on commercial airlines regulated routes and fares. They used a regulated telephone system with ubiquitous wireline stations where calls could be made or received. They watched a handful of stations on regulated television, usually containing three national networks plus some "independents." They bought entertainment through less regulated but nonetheless restricted channels controlled by large record companies. In contrast, people coming of age in the 1970s grew up with new technologies of many kinds. They had cellular telephones for personal use portable music players playing the cassette tapes they had recorded, they used microcomputers, they joined the game culture, and they explored the Internet. In time they learned to rip CDs and download music in digital MP3 files. When the Apple iPod appeared in 2001 they were ready.

Mobile communication is more than just voice communication over mobile telephones. To understand the significance of the broader genre, consider the dramatic shift in telephony from the old Bell System to contemporary cellular infrastructure. The Bell System invented cellular telephony and deployed the first successful analog cellular service in Chicago in 1983. Yet the leadership of AT&T failed to understand the significance of what it had created, giving it away to the RBOCs under the MFJ of 1982. Most of the RBOCs sold off their cellular telephone rights to other entrepreneurs. How did this happen? AT&T was caught in a competency trap from its success with "universal service." The leadership from top to bottom understood universal service as the ubiquitous wireline service model.[8] By mid century the Bell System was about dial tone. Automatic switching required an alternative to the operator-assisted user interface. The answer was dial tone. As late as the 1960s Western Electric and Automatic Electric telephones came equipped with a printed disk in the middle of the rotary dial that had the telephone number for that device and the words "Wait for dial tone." Dial tone meant that the telephone instrument was "in the network" and that universal service was available. Availability of dial tone was the service metric that local exchanges showed to rate-setting public commissions, proving they were doing the job. Restoration of dial tone was key, and local telephone companies boasted of their ability to restore dial tone quickly when it failed. The hallmark of service was that the telephone was "always up."

Neither the telephone nor any part of the infrastructure belonged to the user. The user was reminded of this by the monthly charge for the telephone on the bill. If the user turned the telephone over, a statement on the bottom (sometimes stamped in steel) said the telephone was the property of the telephone company and was not for sale. The telephone instrument was a company-owned station within a company-owned infrastructure. Long-distance calling was by default *station-to-station*; long-distance charges were incurred if anyone at the receiving telephone answered. To ensure that a particular person received the call, the call could be placed *person-to-person* for an additional fee. Person-to-person calls remained operator-assisted long after automatic long-distance dialing was introduced. The company owned radiotelephony. High demand and constrained supply meant that a subscriber sometimes had to wait until a subscription came open before receiving service. No competitors offered faster access. A mobile telephone was based in a vehicle, not carried on the person. Radiotelephone transceivers were large and required the power of a vehicle battery to operate. They were not really portable. Even the early analog cellular telephones were bulky and heavy, and seldom left vehicles. The vehicle was a station to which the user would go for access.

C. Martin Cooper, a key figure in the development of analog mobile telephony in the United States, with an early cellular phone. (Photo from Wikimedia Commons; copyright Rico Shin, reprinted under GNU Free Documentation License 1.2)

Under the regulated monopoly of the Bell System, actual ownership and affective ownership were the same. The telephone company owned the infrastructure, and the user was a subscriber or paying customer (in the case of public telephones, for example). The user never felt like an owner. This was reinforced by the Bell System's multi-decade program of advertising signaling that the Bell System would take care of telephone users. Affective ownership began to shift toward the user when automatic switching came in, removing the calling party from direct contact with telephone company employees. In the 1960s and 1970s more elaborate billing schemes allowed calling parties to make toll-free long-distance calls, punch in codes to access calling points (e.g., AT&T USA Direct for international calls) and direct billing to calling cards, credit cards and other numbers. The calling party was increasingly the operator of the infrastructure, although it was still necessary to go to the station to make the call. When regulated monopoly status disintegrated and competition entered the local and long distance markets, affective ownership changed quickly.

From the first appearance of the analog AMPS in 1983 cellular telephony began growing. By 1988 there were 1.6 million customers in the US, and by 1990 it was clear that the analog infrastructure could not handle growth in demand (Privateline.com, p. 9, n.d.). In 1991 the IS-54B digital standard was adopted for US cellular service (Privateline.com, p. 9, n.d.).[9] Improvements followed steadily, and as of 2009 there were over 275 million cellular phones in service in the US ("List of Countries by Mobile Phone Use," 2010). Users no longer felt tied to their service providers, and the competitive cellular service market further blurred the relationship between the user and the service provider. This portability brought affective ownership closer to the user than ever before. The cellular telephone became "my telephone" for millions of users, whether or not they personalized the instrument or the service with decorations and custom ring-tones. By the 1990s it no mattered little who owned the instrument or the infrastructure behind it. The trade-offs in this shift are shown in Table 3.1.

Table 3.1 Comparison of Bell System Model for Telephony
to Contemporary Cellular

Characteristic	Bell System Model	Contemporary Cellular Model
Nature of telephone	Station to which user goes	Accessory carried on person
Nature of service	Dial tone, "always up"	Sometimes in range, sometimes not
Telephone number	Number is to the phone	Number is to the person

Mobile communication took on broader significance with the advent of portable computers, the spread of wireless Internet, the promulgation of late-generation cellular network capabilities, and the convergence of these onto increasingly sophisticated devices operating on software platforms such as iPhone OS and Android. Apple Computer and Google attempted to shift the user's sense of "telephone" from the network provider—the long-established model of the Bell System—to the maker of the platform that the user carried as a personal accessory. That accessory happened to be a telephone, among other things, and to make telephone calls it connected to an infrastructure owned by someone. However, to the user it did not really matter who owns that infrastructure. Telephony remained an important part of mobile communication, but was increasingly a component in a larger constellation of information services linking individuals to each other and to content. In this respect, mobile communication continued to depart from the regulated monopoly model that prohibited common carriers like the AT&T Bell System from providing information services not strictly related to the telephone infrastructure (e.g., directory assistance).

Conclusion

System, networks and science remain at the heart of the business, as do market liberalization and digital technologies. What happened to consolidation, rationalization, and regulated monopoly? Those no longer seem salient. Local exchange carriers, the descendants of the RBOCs, now seek to eliminate wireline service because it is too expensive (Lasar, 2009). Was the technical and institutional progress of Ma Bell an illusion, occluding from view the greater glory that the unfettered marketplace would have provided? History does not reveal its alternatives. Nevertheless, given how much of the era of mobile communication builds upon the earlier eras, it is worth asking whether there are lessons from this history.

The long period of consolidation, rationalization and regulated monopoly arguably served the functional purpose of stabilizing the emerging market for telephony so that long-term learning could take place. The concept of universal service proved to be politically salient under two different meanings attached to it. Theodore Vail's 1908 concept of universal service—one policy, one system—made sense if the alternative was to subscribe to every service in order to reach every customer. Over time, inter-operation was accomplished through automatic switching, and the Internet later proved that multiple, heterogeneous networks could be linked together through gateways without having a single service provider. But those insights were learned, and would probably not have been technically feasible in the early twentieth century. The first notion of universal service spawned the second, and by the 1930s telephony had become so successful and important that universal service implied a "right of access" by customers. Any telephone company *entitled* to cover everyone in the region without competition *had* to cover everyone in the region. By integrating local telephone systems through long-distance infrastructure, the ability to call anywhere from anywhere became technical reality. The distinction between local and long distance service was for some decades a matter of pricing; the Bell System kept long-distance rates high to cross-subsidize local telephony and keep local rates low. After deregulation the distinction between local and long-distance began to disappear altogether. Today, most cellular services simply charge "minutes" for calls anywhere within a very large geographic region (e.g., the United States).

The regulated monopoly era also created the unified telephone numbering scheme. AT&T led the world in standardizing telephone addressing, first at the local level, then at the national level through direct long-distance dialing and the creation of Number Plan Areas (area codes) in the North American Numbering Plan. AT&T's leadership was evident at the international level through the implementation of direct international dialing and the creation of World Zone 1 (Andeen, 1998). Telephone companies in most other countries followed AT&T's lead, normalizing telephony throughout most of the world so that people from anywhere could use the telephones anywhere without too much difficulty. When the US began to deregulate telephony in the 1980s, many other countries also followed suit.

The AT&T Bell System, as successful as it was under the regulated monopoly model, ultimately reminded the country of the risks of monopoly, even regulated monopoly. Over the decades the regulatory apparatus became co-opted by the entity it was supposed to be regulating, both at the local level of public commissions and at the national level in the FCC.[10] The leadership of AT&T fell into a true competency trap, as well, seeing cellular telephony as little more than a higher-capacity extension of the mobile telephone structure of 1946. Mobile phones, in the minds of AT&T's leaders, were tied to vehicles used by emergency services and other functions that were mobile only as a requirement of their work. AT&T gave away cellular telephony in 1984 and tried to make its fortune in computers. Cellular telephony proved valuable beyond imagination, and the company's descendants eventually bought their way back in at huge cost. In the mean time, AT&T's venture into computers was an abysmal failure (Wikipedia, *AT&T Computer Systems,* 2010).[11] It is doubtful that the AT&T of the 1980s could have understood the revolution in mobile communication, much less facilitated it.

The AT&T Bell System was the most successful monopoly in the history of the world. It had a long list of tremendous accomplishments in the middle of the twentieth century. It created a version of mobile communication through deployment of ubiquitous telephone stations tied to expansive wireline network infrastructure, and it improved that infrastructure so that use became easier as time went on. In the end, the AT&T Bell System proved that a global, user-operated infrastructure was technically feasible and operationally workable. At that point, its job was done.

Acknowledgments

The author is deeply indebted to James Katz and three anonymous reviewers for assistance with this chapter. Errors, omissions, and other foolishness remain the responsibility of the author.

Notes

1 Bowker adopts the broad concept of infrastructure developed by Star and Ruhleder (1996).

2 Subsequent analysis of the Hawthorne Studies led to identification of the *Hawthorne Effect*: that subjects in studies respond to the fact that they are being studied, and not only to the particular treatments being studied (Landsberger, 1958). Landsberger taught at Cornell, long the home of the journal *Administrative Science Quarterly.*

3 The early science fiction of Jules Verne and H.G. Wells was very popular between 1870 and 1910; and the Nobel Prizes were first given in 1901 lending great prestige and romance to science and scientists.

4 See Chandler, Alfred *The Visible Hand: The Managerial Revolution in American Business,* Cambridge, MA: Harvard Belknap, 1977; and Peritz, Rudolph J.R., *Competition Policy in America, 1888-1992: History, Rhetoric, Law.* Oxford: Oxford University Press, 1996.

5 The Civil Aeronautics Board was eliminated altogether during the Reagan administration in 1984, and its few remaining powers were transferred to the Surface Transportation Board.

6 Reagan's defense secretary, Caspar Weinberger, argued that breaking up AT&T would damage US national security. Reagan's attorney general, William French Smith, continued the anti-trust case.

7 An image of this copyrighted cartoon can be readily obtained by typing the words *internet dog* into any browser's search engine.

8 For one view, see King and West, 2002.

9   This was TDMA, time division multiple access; the first of a number of digital standards adopted in the US.
10   A high-level staffer in the FCC explained to the author in the early 1980s, "The FCC doesn't know much about telephones; we let AT&T run the phone system and the commissioners go to lunch with the broadcasters."
11   It was not as though AT&T lacked expertise in computers: Bell Laboratories had invented the transistor and had created the Unix operating system. What AT&T lacked was the knowledge of how to compete in the computer business.

References

"Airline Deregulation Act." (2010, February 7). *Wikipedia.* Retrieved March 23, 2010 from: http://en.wikipedia.org/wiki/Airline_Deregulation_Act.

"AT&T Computer Systems." (2010, February 3). *Wikipedia.* Retrieved March 29, 2010 from: http://en.wikipedia.org/wiki/AT&T_Computer_Systems.

"DARPA." (2010, March 30). *Wikipedia.* Retrieved March 23, 2010 from: http://en.wikipedia.org/wiki/DARPA.

"Letter to the Attorney General from the American Telephone and Telegraph Company Outlining a Course of Action Which It Has Determined Upon; The Attorney General's Reply; The President's Letter to the Attorney General." (1914). Washington, DC: Government Printing Office.

"List of countries by mobile phone use." (2010, March 25). *Wikipedia.* Retrieved March 23 from: http://en.wikipedia.org/wiki/List_of_countries_by_number_of_mobile_phones_in_use.

"Morrill Land-Grant Colleges Act." (2010, March 23). *Wikipedia.* Retrieved from: http://en.wikipedia.org/wiki/Morrill_Land-Grant_Colleges_Act.

"Sociotechnical Systems." (2010, March 10). *Wikipedia.* Retrieved from: http://en.wikipedia.org/wiki/Sociotechnical_systems.

"Spacetime." (2010, March 10). *Wikipedia.* Retrieved from: http://en.wikipedia.org/wiki/Spacetime.

Andeen, Ashley and King, John L. (1998). "Addressing and the Future of Communications Competition." *Information Polity*, 6, No. 1, 17-46.

AT&T. (2010). *Milestones in AT&T History.* Retrieved from: http://www.corp.att.com/history/milestones.html.

Bowker, G. (1994). *G.C. Science on the Run: Information Management and Industrial Geophysics at Schlumberger, 1920-1940.* Cambridge: MIT Press.

Brooks, John. (1980). *Telephone: The First Hundred Years.* New York: Harper and Row.

Brown, D.C. (1980). *Electricity for Rural America.* Westport, CT: Greenwood Press.

Brown, Merrill. (1981, April 9). Weinberger asks Justice Dept to drop its suit against AT&T. *Washington Post,* p. B10. Retrieved from: http://proquest.umi.com.

Carey, James W. (1989) "Technology and Ideology: The Case of the Telegraph," *Prospects*, 8, 1983: 303-25.

Cybertelecom. (2010, March 26). *Telecommunications act of 1996.* Retrieved from: http://www.cybertelecom.org/notes/telecomact.htm.

Fors, Geoff. (2010, March 17). *The mobile telephone in Bell System service, 1946-1985.* Retrieved from: http://www.wb6nvh.com/Carphone.htm.

Freeman, Chris and Soete, Luc. (1997). *The Economics of Industrial Innovation.* Cambridge, MA: MIT Press.

Garber, Steve. 2010, January 28. *The Decision to Go to the Moon: President John F. Kennedy's May 25, 1961 Speech before a Joint Session of Congress.* Retrieved from: http://history.nasa.gov/moondec.html.

Graham, Laurel D. (1999). "Domesticating Efficiency: Lillian Gilbreth's Scientific Management of Homemakers, 1924-1930." *Signs,* Vol. 24, pp. 633-675.

Grodinsky, Julius. (1930). *Railroad Consolidation: Its Economics and Controlling Principles.* New York: Appleton.

Harvard College. (2010). *The human relations movement: Harvard Business School and the Hawthorne experiment (1924-1933).* Retrieved from: http://www.library.hbs.edu/hc/hawthorne/intro.html#i.

Hedstrom, Margaret and King, John L. (2006). "Epistemic Infrastructure in the Rise of the Knowledge Economy. In Kahin, B. and Foray, D. (eds.), *Advancing Knowledge in the Knowledge Economy.* Cambridge, MIT Press, pp. 113-134.

Israel, Paul. (1992). *From Machine Shop to Industrial Laboratory: Telegraphy and the Changing Context of American Invention, 1830-1920.* Baltimore: Johns Hopkins University Press.

John F. Kennedy Presidential Library and Museum. (1961, May 25). *Special Message to Congress on Urgent National Needs,* p. 4, Retrieved from: http://www.jfklibrary.org/Historical+Resources/Archives/Reference+Desk/Speeches/JFK/Urgent+National+Needs+Page+4.htm.

King, John L. and Joel West. (2002). "Ma Bell's orphan: US cellular telephony, 1947-1996." *Telecommunications Policy, 26,* Nos. 3-4, 189-203.

Landsberger, Henry. (1958). *Hawthorne Revisited,* Ithaca: Cornell University Press.

Lasar, Matthew. (2009, December 30). *AT&T: Landline phone service must die; only question is when.* Retrieved from ArsTechnica: http://arstechnica.com/telecom/news/2009/12/att-landline-phone-service-must-die-only-question-is-when.ars.

Mosca, Manuela. (2006) On the Origins of the Concept of Natural Monopoly (December). *Università di Lecce Department of Economics Working Paper* No. 92/45. Retrieved from: http://ssrn.com/abstract=975461.

Nonnemacher, T. (Vol. 25, 2005). Network Quality in the Early Telegraph Industry. *Research in Economic History,* 61-82.

Osterbrock, Donald E., Gwinn, Joel A., and Brashear, Ronald S. (1993). "Edwin Hubble and the Expanding Universe." *Scientific American* Vol. 269, No 1, pp. 84-89.

Privateline.com. (n.d.). *Mobile Telephone History.* Retrieved from Privateline.com: http://www.privateline.com/PCS/history9.htm.

RsX Enterprises. (2010). *World Zone 1.* Retrieved from: http://www.rxs-enterprises.org/CountryCodeDecoder/world-zone-1.aspx.

Star, Susan Leigh and Karen Ruhleder (1996). "Steps Toward an Ecology of Infrastructure: Design and Access for Large Information Spaces," *Information Systems Research,* 7, No. 1, pp. 111-134.

Sterling, Christopher H., Bernt, Phyllis and Weiss, Martin. (2006). *Shaping American Telecommunications: A History of Technology, Policy and Economics.* Mahwah, NJ: Erlbaum.

Taylor, Frederick W. (1911). *The Principles of Scientific Management.* New York: Harper and Row. Retrieved from: http://www.gutenberg.org/etext/6435.

Weber, Max. (1978). *Economy and Society,* (Roth, Guenther and Wittich, Claus, Eds.), Berkeley: University of California Press.

Yates, JoAnne. (1989) *Control Through Communication: The Rise of System in American Management,* Baltimore: Johns Hopkins University Press.

4

Sustainable Early Warning Systems: HazInfo Sri Lanka

Gordon A. Gow and Nuwan Waidyanatha

Early in the morning on December 26, 2004, seismic equipment reporting to the Pacific Warning Tsunami Center (PTWC) in Hawaii detected a massive earthquake off the coast of northern Sumatra. Scientists working at the Center soon determined that a tsunami threat existed for the Indian Ocean and concern began grow about the need to inform people living in the region. While outside the immediate jurisdiction of the PTWC, scientists there nevertheless decided to post a bulletin on the Center's public website at 02:04 Zulu time (08:04 local time in Sri Lanka) warning of the possibility of a tsunami in the region.

Minutes after the earthquake, places near the epicenter like Banda Aceh saw the first waves arrive. However, in locations like Kalmunai and other areas along the east coast of Sri Lanka it took 90 minutes or more before the initial tsunami reached the shoreline. Then, for another hour, several more tsunamis inundated the region, destroying fishing villages and tourist resorts throughout the basin. A quarter of a million people would eventually perish—over 40,000 in Sri Lanka alone. In hindsight, many now agree that the major failure of that day was not in the *detection* of a potential tsunami—the scientists at PTWC had issued a public bulletin—but rather in the ability and/or willingness of national and local authorities to *disseminate* a warning to local populations. As noted, many communities and resorts along the coastline had ample time to take action and evacuate at-risk areas well before the tsunami arrived, had only they been given that decisive bit of information that resided on the PTWC website.

A central lesson from the 2004 Tsunami is that improvements to hazard detection in the Indian Ocean and elsewhere may be necessary but they are only one half of a comprehensive approach to more effective public warning systems. The account of the unheeded PTWC bulletin suggests that a more fundamental challenge going forward will be to establish—and to sustain—socio-technical systems that will link the global information infrastructure with the means to alert local populations at the community level, or what is sometimes termed the "last mile."

As we will argue in this chapter, mobile phones can play a vital role in building sustainable last-mile early warning systems provided that they are deployed within an effective use strategy. Our argument is based on empirical observations drawn from the HazInfo Sri Lanka project—a multi-year action research study that examined the role of information and communication technologies in supporting last-mile early warning. Those observations are then considered in light of a theoretical framework that blends the concept of "effective use" from the community informatics literature with Ling's theory of mediated ritual interaction. The framework advances our understanding of the sustainability challenge by offering insights as how the mobile phone can contribute to social policy in support of local early warning capability through community participation.

Mobile Phones and Last-Mile Warning

In spite of all the initiatives and attention paid to early warning systems since the 2004 Sumatra tsunami, there remains a persistent challenge in building and sustaining local warning capability at the last-mile. The last-mile refers to the final stage in a warning system that connects individuals within local communities to urgent risk information, such as a tsunami warning. Often the vulnerability of a local community to risk can be mitigated if there is a capability to reach people at the last-mile with timely alerts or warnings, enabling them to take action to save lives and protect property. Specialized technical systems such as pole-mounted sirens are effective in some cases but there is ongoing uncertainty as to the level of public understanding of and response to these kinds of single purpose systems. Research on tsunami sirens in Hawaii, for example, suggests that even when they have a long history in the community, a high level of public awareness does not "equate with increased understanding of the meaning of the siren, which remains disturbingly low" (Gregg et al., 2007). Having examined decades worth of research on human response to warning messages, a leading expert in the field has concluded that "[a] siren—like any other noise—is not a warning; at best it may alert some, but many will ignore it" (Drabek, 2010, p. 60). Moreover, even if siren systems are proven effective in some cases, maintaining specialized technology, particularly in remote areas, can be a costly undertaking for low probability events like tsunamis and may simply be deemed uneconomical in certain settings.

By contrast, major media sources including radio and satellite or cable TV systems are often distributed from centralized head-ends located far away in major urban centers, usually with a limited capability for cutting into programming with information messages targeted to smaller geographic segments within their larger footprint. Nonetheless, blanketing large areas with region-wide alerts could be (and is) done in the name of public safety, but over the long term it runs the risk of desensitizing local communities to these alerts, as many of the messages may not be immediately relevant to them. Findings from research on human response to warnings suggests that well-targeted messaging is an important consideration in maintaining the credibility of an emergency alerting system over time (Partnership for Public Warning, 2003).

In one sense, then, the challenge in developing countries like Sri Lanka is not unlike that in the rural and remote parts of developed countries like Canada. Local warning capability is limited by access to communications technology and the ability to reach populations with localized alerts (LIRNEasia, 2005). However, in the case of developing countries the problem is somewhat more acute inasmuch there are serious challenges in terms of establishing a reliable source of local warning messages in the first instance. In some cases this stems from political situations that systemically hinder the development of reliable local warning information (Samarajiva, 2005).

The challenge of establishing a reliable last-mile warning system is threefold: (1) how to reach individuals with urgent messages on a 24/7 basis—especially late night/early morning hours; (2) how to establish a system that will target populations with local warnings that are specific for their community while avoiding inundating them with non-relevant messages; and (3) how to sustain the system over the longer term, particularly when sources of funding are limited or uncertain.

In light of these requirements, the mobile phone is a promising technology. As an "always-on" personal communications device, the mobile phone would at first blush appear to be an emergency alerting tool par excellence. Alerts can be targeted to specific individuals using SMS or to specific geographic locations with the use of cellular broadcast (Wood, 2006). The mobile phone is also a form of social media that enables peer-to-peer warning practices to evolve where there is otherwise uncertainty as to reliability of official sources of information. Drabek has observed, for instance, that unofficial warning through friends and family members is a significant and effective phenomenon during emergencies (Drabek, 2010, p. 52). Moreover, the mobile phone is a communications tool that tends to be integrated into the everyday communicative practices among individuals, thereby increasing the likelihood that it will be maintained and functioning should an alert be issued. In other words, the mobile phone has the potential to respond to the first two requirements noted above. The next two sections describe the HazInfo project in Sri Lanka and a set of observations emerging out of that project that confirm a vital role for mobile phones in sustainable early warning systems but also reveal that the goal of sustaining a system over the long term remains somewhat more elusive.

The HazInfo Project in Sri Lanka

From late 2005 to mid 2008, the authors were involved in a research project in Sri Lanka intended to address concerns related to last-mile warning. The HazInfo Project was made possible with funding from Canada's International Development Research Centre (IDRC) and headed by the policy and regulation capacity-building organization LIRNEasia, along with support from several local organizations including Sarvodaya, the largest and most established NGO in Sri Lanka. A primary aim of the project was to establish and evaluate a community-based hazard warning system that would act in concert with any initiative that the national government might introduce in future. The system used a range of wireless technologies including but not limited to mobile phones.

The first step was to establish a reliable source of early warnings for tsunamis and other hazards. Sarvodaya introduced a "Hazard Information Hub" at its Community Disaster Management Centre in Moratuwa, where volunteers were recruited to monitor various websites on a 24/7 basis for "events of interest" that might be cause for concern (e.g., an earthquake off the coast of Indonesia). From here, information bulletins would be issued to individuals within 32 participating communities. These individuals—referred to as "ICT Guardians" by the project team— were equipped with one or more wireless communications devices that they were to maintain in good working order at all times. The devices introduced by the HazInfo project included mobile phones, CDMA fixed wireless handsets, addressable satellite radios donated by a corporate partner, and specialized GSM-based Remote Alarm Devices designed by engineers at the University of Moratuwa. (The HazInfo inventory was comprised of several desktop PCs located at the Hazard Information

Hub, connected to the Internet by means of a 128 Kbps microwave link provided by Dialog Telecom. The field component was comprised of 10 mobile phones (Nokia 6600) with service provided by Dialog Telecom; 8 CDMA fixed wireless handsets, with service provided by Sri Lanka Telecom; 56 Worldspace satellite radios with the Disaster Warning Response and Recovery service provided by Worldspace. (A transponder channel was also made available to Sarvodaya on a temporary basis for broadcasting news and information to its villages); and 15 GSM Remote Alarm Devices donated by the University of Moratuwa.)

When an ICT Guardian received an information bulletin over one of their devices, they were instructed to take action based on training provided to them by staff from Sarvodaya. Depending on the nature of the event, this action could range from simply informing the community about a potential risk, to initiating an immediate evacuation. In this respect, the HazInfo was not a "public warning" system in the strictest sense, but instead a social network of designated first responders who would in turn alert their communities through other means, including loudspeakers, temple bells, and word of mouth—a two-step flow, as it were.

Despite various technical issues that were encountered at various stages during the project, a series of exercises conducted over the course of a year (as well as a genuine tsunami alert issued on September 12, 2007) have provided evidence to suggest that a community-based initiative like this can improve the supply of local warnings, even with minimal support from the national government (LIRNEasia, 2008).

However, and perhaps not unexpectedly, the local warning capability established under HazInfo began to diminish when the project came to an end. Local communities are now more aware of the tsunami hazard, but both the communication links and the general state of readiness in these communities have declined significantly in recent months. At the same time there remain unanswered questions as to the reliability of country's official public warning system. Events surrounding the September 12, 2007 tsunami warning in Sri Lanka reinforce the view held by some that numerous practical and procedural matters remain to be resolved before the national system can be considered reliable (Samarajiva, 2007).

Lessons Learned from HazInfo

One of the key objectives set out by HazInfo project team was to support the integration of the various communication tools into the everyday activities of the community. The view held by the project team was that integration is essential to creating long-term demand for the system and to support ongoing response readiness for future emergency incidents.

Yet by looking at the results achieved, it became clear that there were significant barriers to integration. On the one hand, the expectation for the satellite radios was that they would be reasonably well integrated into community life because they also offered access to a variety of daily information and entertainment programming to the communities. (Worldspace in fact made a transponder channel available to Sarvodaya for its news and information service to be broadcast to the communities over the basic tier of the AsiaStar WS satellite. The use of this channel is, however, temporary and it was reported to me that it is to cease.) Technical challenges faced by users within the communities played a role in reducing this outcome. On the other hand, the expectation for the Remote Alarm Devices was less certain because they were highly specialized devices designed largely to perform a single function. As it turned out, these devices are more likely to be disregarded by community members in part because there is otherwise little call for them to be used on a regular basis.

However, the villages that were given access to mobile phones and wireless CDMA handsets appear to have made the most progress in terms of integrating these devices into the daily life of the community. In fact at one point, Sarvodaya had to take steps to constrain the use of the deployed mobile phones after it became apparent that the cost of calls was exceeding the allocated budget. The "problem" was that individuals in the communities were using the phones more frequently that was expected—typically to make personal calls. Although the mobile phones did present some minor technical issues in terms of their Java-based alerting feature, the fact is that they were the most actively used and integrated technology deployed within the project.

The lesson here is important: in contrast to the satellite radios and Remote Alarm Devices, the mobile phone offered personal access to peers, to chat and exchange information, and to maintain important social connections—albeit sometimes unrelated to the intended purpose of hazard warning. With that possibility open to them, it seems plausible that the ICT-Guardians and others in the community were far more likely to want to use the phone on a daily basis, to want to take care of it by keeping batteries charged, to want to keep the phone turned on at all times in anticipation of incoming calls, and to want to be more aware of the functional capabilities (and limitations) of the device.

Based on this observation, one might be tempted to argue that the sustainable solution for local hazard warning is simply to supply every village with access to a telephone or mobile phone. Indeed, the case of an Indian village in Pondicherry, saved from the tsunami by a phone call from a concerned relative in Singapore, validates this to a certain extent (Muthalaly, 2005). However, we must also be mindful of the fact that thousands of locals and tourists in other places were caught in the tsunami, many of whom did have a mobile phone, thereby suggesting that access to a technology is not enough on its own to provide assurance that people will be alerted during an emergency. Even with the mobile phone, social networks are necessary to circulate information in a timely manner if people are to be alerted.

Nonetheless, in the face of wide variety of solutions available for public warning, the findings concerning the mobile phones deployed for the HazInfo Project reveal one very important relationship in terms of access: give individuals a tool that they want to use, keep it simple, and they are more likely to use it, maintain it, and even possibly experiment with it. In other words, basic communication capabilities are more likely to be sustained if individuals in the community take a personal interest in them. Moving beyond access to achieve everyday integration of last-mile technology into the community is therefore a key consideration in working toward a social policy that supports sustainable early warning systems. The mobile phone can play an important role in this regard provided that it is incorporated into a strategy that enables and promotes the sharing of local risk information.

Going Beyond Access to Achieve Effective Use

A theoretical model that goes beyond access to consider the broader social context for ICT adoption is necessary if we are to develop an actionable strategy that can lead to reliable and sustainable local warning systems. In this respect, the concept of effective use is helpful. Michael Gurstein, a leading thinker in the area of community informatics, adopted the term in an effort to define a more holistic, participatory approach for studying and encouraging ICT adoption:

> The ongoing process of seeing the DD [digital divide] only in terms of "access" further aggravates and perpetuates the notion that with an ICT platform there will be a relatively small number of producers and a very large, even universal, set of consumers. Meanwhile, of course, the technology is such as to allow for each to be both a consumer and a producer of information and ... productive knowledge-intensive goods and services within an electronically enabled environment. ... The challenge thus, is to ensure not simply "access" but "effective access" or "use," that is, access which can be used and made effective to accomplish the purposes that individuals might set for themselves (Gurstein, 2003).

This distinction between passive consumers and active producers of information is an important consideration in light of the massive investment in sophisticated tsunami early warning systems that seem to provide little opportunity for community's themselves to become directly involved in the management of local risk knowledge. Along these lines, the concept of effective use appears again in Gurstein's observations about the 2004 Tsunami, noting that the disaster was a prime example of the gap between access and effective use. He reminds us that information about the hazard was available but that individuals and communities in harm's way had little or no ability to make use of it. He contends that in this case "as elsewhere, it is the 'social' organization of the Last Mile which will mean whether the information is used or not and whether lives are or are not saved" (Gurstein, 2005, p. 16).

Gurstein argues that an effective use response to the 2004 Tsunami would place more emphasis on developing local capabilities to manage and use information that is currently available and, moreover, to develop community-based networks for dissemination and emergency response (2005, p. 17). Similar to a 2005 UN report that identified "people-centred" early warning as a future priority (United Nations, 2006), Gurstein contends that a long-term strategy must seek to develop and integrate local knowledge by cultivating extended communities of practice linked through ICTs and shared social arrangements. Similar views are now commonplace among those in the disaster research community (National Research Council, 2006).

Mobile Phones and Local Risk Knowledge as Ritualistic Interaction

In certain respects the HazInfo Project was well aligned with an effective use strategy as defined by Gurstein. In fact, the project organizers had recognized from the outset that access to the technology was only a pre-condition for the long-term success of the system. To the extent that HazInfo established a basic administrative structure and provided training to the participating communities it achieved a modest level of success in going beyond access to promote effective use of technology. However, in terms cultivating an extended community of practice around local risk knowledge, the project was never able to achieve significant gains; this despite best efforts on the part of the project team to encourage local interaction by planning for a communications network that would eventually function with two-way interactions between the Hazard Information Hub and local community members.

Further empirical study is needed to understand how the enthusiastic uptake of the mobile phone as observed in the HazInfo project can be capitalized upon to foster a sustainable community of practice around local risk knowledge and, in turn, improve local warning preparedness on a long term basis.

Recent theoretical work suggests a promising line of inquiry in this regard. Ling has argued, for example, that the mobile phone can serve to reinforce "social coherence" through the ritualistic elements it engenders within small groups (Ling, 2008b). Ling's notion of ritual is an amalgam drawn from various sources, offering a plausible approach to fostering a community of practice through mediated peer interaction:

> [Ritual] … involves the establishment of a mutually recognized focus and mood among individuals, and it is a catalyst in the construction of social cohesion. The focus is not on obsessive or repetitive behaviour, although ritual interactions can take place in these settings. Rather, the emphasis is on a group process and the outcome of that process (Ling, 2008b, p. 9).

A potentially significant line of inquiry emerging out of this claim is how social policy can work to encourage the formation of ritualistic patterns around the sharing of local risk knowledge such that it becomes integral to everyday communications among the ICT-Guardians within these communities. One vehicle for doing so could take the form of community-based risk mapping that uses mobile phones to collect and share data. Along these lines, a recent study reported by Tran, et al., on the use of GIS technology for flood risk mapping in Vietnam offers some interesting possibilities for adaptation by using mobile phones as tools for community members to produced and share risk information (Tran et al., 2008).

Tran and his collaborators share the view that "communities have shown themselves to be a source of strength, contributing innovative ideas and local knowledge which, when mobilized and used appropriately, can lead to solutions that can make a fundamental contribution to mitigating the negative impacts of natural disasters." Building from this premise, they conclude that "the most successful way to do this is to engage in a process that enables local knowledge to be transferred from the mind to the map." The mobile phone can be a valuable tool in this regard, especially when combined with the power of a digital camera and GPS functionality. ICT-Guardians could be tasked with identifying and sharing information about local risks through both visual and textual methods as a form of "public authoring" (Proboscis, 2008) within the context of a hazard mitigation initiative. The aim of such a project would be to promote community participation in capturing and representing local knowledge on shared maps. A key finding from Tran's study is promising in terms of this potential:

> This flood risk mapping successfully transferred unrecorded local knowledge into maps. The process of developing risk maps also mobilized the participation of the local population and succeeded in establishing trust, respect and an exchange of information among local communities and local authorities as well as local planners. This involvement assisted enormously in the development of a safer community plan.

Moreover, the results also suggest that participants experienced a level of "mutually recognized focus and mood," leading to greater commitment to the process and its outcomes:

> Another experience from the mapping process showed that villagers have subsequently become more aware of their risks. Incorporating existing and traditional disaster coping mechanisms of the community into the disaster management plan increased the plan's acceptance among villagers and ensured an independent commitment. Once plans have been implemented, farmers feel responsible for their involvement, since they drafted the plans themselves. This reduced the costs of external monitoring and ensure the long-term sustainability of the approach.

The authors note that "community spirit" is a vital force in maintaining motivation, and that this may need to come from local boards or other organizations with roots in the community. By putting the data gathering capabilities into the hands of local community members with support from a coordinating body and open GIS platform, perhaps in conjunction with a community group such as Sarvodaya's Disaster Management Centre, the mobile phone might prove to be a cost effective means of generating and sharing risk knowledge. A working example of this proposed approach can be seen in the Ushahidi online platform that has demonstrated the practical use value of crowdsourcing to support crisis management specifically within developing countries. Ushahidi is a free and open source (FOSS) geospatial platform "that allows anyone to gather distributed data via SMS, email or Web and visualize it on a map or timeline. Our goal is to create the simplest way of aggregating information from the public for use in crisis response" (Ushahidi, 2010). A similar approach based on the Ushahidi model might be suitably adapted for risk mapping and emergency preparedness in an effort to mitigate crisis situations at the outset.

Risk mapping could be an initial focus for the initiative but the extended benefit for public alerting, as well as reducing community vulnerability to natural disasters overall, would derive from the social capital and "interaction ritual" that emerges as a result of ongoing interactions among community members on risk and risk-related topics. Ling, for example, suggests that a process of entrainment might unfold when routine patterns of telephone contact between colleagues "result in a type of solidarity …. and revitalization of group identity" (Ling, 2008a, p. 172).

Risk awareness and, ultimately, community disaster preparedness might thereby be enhanced within the community in part because members become more informed, but also because of the social capital created and sustained through the use of mobile phones in contributing to the risk-mapping project. In this sense, the local community designates working on the risk knowledge project would act as "weak ties" or "bridging capital" that connects local communities to each other and to the coordinating Disaster Management Centre.

Insofar as funding constraints remain a key consideration in countries like Sri Lanka, the mobile phone offers a relatively cost-effective tool that reduces transaction costs in specialized group forming (Benkler, 2006; Shirky, 2008). Moreover, the perceived benefit of participation in a hazard information network might be further enhanced if members of the community were also to recognize the value of risk knowledge sharing to support decision-making around commonplace activities involving household and community planning. In this way the investment in a mobile phone also serves to support a social network that could reinforce local emergency preparedness within wider effective use strategy sustained through ongoing development related initiatives such as health services and local business projects (Mecheal, 2008; Overa, 2008).

Acknowledgments

The authors acknowledge International Development Research Centre, LIRNEasia, and Sarvodaya for the research opportunities they made possible through their leadership in organizing and supporting the HazInfo Sri Lanka initiative.

References

Benkler, Y. (2006). *The Wealth of Networks: How Social Production Transforms Markets and Freedom*. New Haven: Yale University Press.

Drabek, T.E. (2010). *The Human Side of Disaster*. Boca Raton: CRC Press.

Gregg, Chris E., Houghton, Bruce F., Paton, Douglas, et al. (2007). Tsunami Warnings: Understanding in Hawai'i. *Natural Hazards* (40), 71-87.

Gurstein, Michael. (2003). Effective use: A community informatics strategy beyond the Digital Divide. *First Monday*, 8 (12).

Gurstein, Michael. (2005). Tsunami Warning Systems and the Last Mile. *The Journal of Community Informatics*, 1 (2), 14-17.

Ling, Richard. (2008a). The Mediation of Ritual Interaction via the Mobile Telephone. In J. Katz (Ed.), *Handbook of Mobile Communication Studies* (pp. 165-176). Cambridge: MIT Press.

Ling, Richard. (2008b). *New Tech, New Ties*. Cambridge, MA: MIT Press.

LIRNEasia. (2005). National Early Warning System Sri Lanka: Concept Paper. Retrieved August 10, 2008. Available http://lirneasia.net/projects/2004-05/national-early-warning-system/.

LIRNEasia. (2008). *Evaluating Last-Mile Hazard Information Dissemination (HazInfo): Final Technology Report to IDRC*.

Mecheal, Patricia. (2008). Health Services and Mobiles: A Case from Egypt. In J. Katz (Ed.), *Handbook of Mobile Communication Studies* (pp. 91-105). Cambridge: MIT Press.

Muthalaly, Shonali. (2005, Jan. 1). A phone call saved an entire village. *The Hindu*. Retrieved August, 2008. Available http://www.hindu.com/2005/01/01/stories/2005010107320100.htm.

National Research Council. (2006). *Facing Hazards and Disasters: Understanding Human Dimensions*. Washington, DC: The National Academies Press.

Overa, Ragnhild. (2008). Mobile Traders and Mobile Phones in Ghana. In J. Katz (Ed.), *Handbook of Mobile Communication Studies* (pp. 43-54). Cambridge: MIT Press.

Partnership for Public Warning. (2003). *A National Strategy for Integrated Public Warning Policy and Capability* (PPW Report 2003-01). McLean, VA.

Proboscis. (2008). Urban Tapestries: Public Authoring and Civil Society in the Wireless City. Retrieved June 1, 2009. Available http://urbantapestries.net/.

Samarajiva, Rohan. (2005). Mobilizing information and communications technologies for effective disaster warning: lessons from the 2004 tsunami. *New Media and Society*, 7 (6), 731-747.

Samarajiva, Rohan. (2007, Sept. 20). Tsunami warning tower fails on September 12. *LIRNEasia*. Retrieved August, 2008. Available http://lirneasia.net/2007/09/tsunami-warning-tower-fails-on-september-12th/.

Shirky, Clay. (2008). *Here Comes Everybody: The power of Organizing without Organizations*. New York: Penguin.

Tran, Phong, Shaw, Rajib, Chantry, Guillaume, et al. (2008). GIS and local knowledge in disaster management: a case study of flood risk mapping in Vietnam. *Disasters*, 33 (1), 152-169.

United Nations. (2006, March 27-29). Global Survey of Early Warning Systems. Paper presented at the Third International Conference on Early Warning, Bonn, Germany.

Ushahidi. (2010, Jan. 12). Ushahidi: Crowdsourcing Crisis Information. Retrieved Jan. 12, 2010. Available http://www.ushahidi.com/.

Wood, Mark. (2006). Introduction to Emergency Cell Broadcasting (video). *Cellular Emergency Alert Service Association*. Retrieved June, 2009. Available http://www.ceasa-int.org/library/Intro_final.mov.

5

Mobile Communication and the Environment

Rich Ling and Nisar Bashir

The mobile phone has become a well-recognized part of everyday life around the world. In the developed world, there are often more subscriptions than there are people in a given country and in the developing world they are quickly spreading (ITU, 2007). The mobile phone has made many contributions to modern society and it has also had some less than desirable effects (Ling & Donner, 2009). The mobile phone has also had broader impacts on society. It has changed the way we exchange time-sensitive information.

Mobile phones also have the ability to facilitate the reduction in energy consumption. There are, however, environmental issues with mobile phones that need to be addressed. This chapter will examine these issues.

On the one hand mobile communication in its different forms can contribute to a more energy efficient and greener environment. We will examine the use of mobile communication in traffic management, in developing countries and in the area of digital distribution. At the same time, the mobile phone is not environmentally neutral. The production, use, and disposal of mobile phones each represents a environmental challenge. We will also look at the question of energy consumption and issues of production and recycling of mobile telephone equipment. As the processing power and the capabilities of mobile phones grow, they also produce a larger impact on the environment. In the conclusion we will outline broad policy issues that should be considered.

Mitigating Greenhouse Gas Emissions in Other Industries

Smart Cars/ Networked Roads

Traffic congestion is a nuisance and it is also costly for society both in terms of lost production but also due to the emission of greenhouse gasses. Schrank and Lomax have estimated that in large US cities the average driver is delayed more than 60 hours a year and in these delays he or she wastes more than 36 gallons of fuel (2007). Translating this to CO_2 output, this is more than 300 kilos of CO_2 per driver. Thus, there is a clear environmental incentive to eliminate congestion (Barth et al., 2007). It is possible to think of a series of condition and event detection sensors in the roadway and also inside the automobile that can assist the driver and help to safely route the car through different situations (ITU, 2008, 2007).

Mobile based information distribution has the potential to contribute to a more efficient transportation system. Given the recent commitment to rejuvenate national infrastructure in the US (read: the highway system) and the potential provided by the extreme retooling of the automobile industry, it is possible to think of the car based transport system somewhat like the Internet, i.e., a bunch of packets moving around in a network (Gripsrud, 2009). If the network is blocked or down in one path, there is a quick recalculation of another route. It is clear that there is a flexibility associated with the Internet that is not necessarily available for the highway system. If there is a bottleneck on the Internet, the packets being sent from Los Angeles to say Detroit might be re-routed via Tokyo. The highway system does not have the same flexibility.

Another issue is that driving is often a mundane activity. For instance, we commute along a similar route on a daily basis, and are likely aware of the potential areas of congestion along this route. We may also know about alternative routes, but these are often thought to be less attractive. We do not, however, have an overview of emerging situations (Horvitz et al., 2005). Thus, warnings as to the road conditions ahead and the alternatives available would allow for an optimization of the transport system.

There is generally the assumption that the most fuel-efficient traffic pattern would be to route vehicles to those roads where there is the least congestion. The idea is that the least CO_2 would be generated when there is the freest flow of traffic. However, Barth et al. suggest that this is not always the case (2007). They outline a situation where sending vehicles to the more open highway reduces the travel time, but it also generates more fuel consumption than were the traffic to be routed to a highway where there is free flow but at a more moderate speed. This points to the issue of optimal density and also points out the tension between individual drivers wishing to use the fastest route versus the broader social goal trying to manage the diseconomies of the transport system while not unduly hindering the flow of traffic. The tension between individual desires and the public good are not new in the area of transport. As noted in the footnote above, informal forms of road information (i.e., warning other drivers about speed traps etc) has been a part of transportation since the time of CB radios. Indeed in some cases this technology is still in use (White & White, 2009). Some drivers have been motivated to contribute road-based information to other drivers, in these cases in order to circumvent authorities via, for example SMS notification. There is a basic question as to whether drivers would be willing to contribute to more centralized systems if there was the sense that it was hindering their travel.

Mobile traffic has been used in various contexts to follow the ebb and flow of traffic. Orange, for example has used it to map the movement of people during World Music Day (http://www.urbanmobs.fr/en/france/) and the Media Lab has developed the "Real time Rome" application, which traces individuals' movements as they move about the city.

If cars were equipped with GPS able to transmit its location, then it would be possible to calculate stoppages. Further, with the addition of an onboard display, the real-time information could be sent back to the vehicles. The mobile network can be used to carry both monitoring information to highway authorities, and from the authorities back to cars with GPS positioning systems. Cars and trucks could be enabled with various chipsets allowing them to communicate over the Internet.

There are many approaches to this issue. The thing that is common to many of the approaches is that dynamic real-time information gathered from a net-wide analysis is fed back into each automobile and displayed on a frequently updated display, alternatively a mobile phone (Yang & Wang, 2007; Nadeem et al., 2004). In some cases, the systems suggest using a centralized system for the gathering of data from all vehicles on the road. This is then examined in a central data collection/analysis point, potential areas of congestion are identified and then these are sent back out to the vehicles that are on the point of entering congested areas. These vehicles are then alerted to the system ahead and alternative routes are suggested (Horvitz et al., 2005).

In summary, real time overview of traffic congestion and the alternative routing suggestions can help to optimize the transportation system. This can be done by directing traffic away from congestion. That is there will be more efficient roadway use and more evenly flowing traffic, less fuel consumption and less CO_2 emission.

Micro-Coordination

In the previous section we examined the role of mobile communication in the "tuning" of the broader transportation system. In this section there is also a focus on the interaction between transportation and mobile communication. Where in the previous discussion the individual was simply informed of the best route through a particular section of the highway system, this section will examine the way that mobile communication is used by individuals to work out their daily affairs.

The analysis begins with the idea that mobile communication makes us individually addressable. When using a mobile phone, in most cases we are calling to a specific individual where with landline phones we were calling to a physical place in the hopes that our intended interlocutor would be nearby. Now, instead of "coming to the phone," the phone is with us, regardless of where we are (Ling & Donner, 2009). This individualizes communication. In many cases, we no longer call to a home or a central receptionist; we call directly to a particular individual.

The mobile phone makes the planning process more individualized and direct. Urban development has been so complex, diversified and diffuse that there is an increased need for coordination (Rasmussen, 2003; Gripsrud, 2009). There is less reliance on calls to geographic locations and on external coordination systems such as time (Ling, 2004) giving us the possibility to make each trip more efficient since travel can be redirected as new exigencies develop. This has been referred to as micro-coordination (Ling & Yttri, 2002).

In some ways, this functionality has been a part of control and dispatch associated with operators commercial and public vehicle fleets for some time (Hsiao et al., 2008). The dispatching function has allowed for more efficient fleet management, better routing and better load management (Jolley, 2006). New technology has challenged the more central dispatch model by allowing individuals to sidestep the more centralized control of the dispatcher (Manning, 1996; Townsend, 2000; Hsiao et al., 2008). Townsend, for example, describes the way that taxi drivers use the mobile phone to go around the more centrally dominated system of dispatchers and to alert their colleagues as to where their services are needed.

Applying this model to the private sphere, it is possible to assert that individual addressability can facilitate better coordination of people as they move about (Ling & Haddon, 2003). It is, possible to suggest that this potential will generate the number of trips since people will plan and execute travel that otherwise would not have been made (Gripsrud, 2009).

Digital Distribution of Reading Material

Another area where digital and wireless technologies can make an impact on the environment is in the distribution of reading material. There is a high cost associated with the packaging and delivering physical objects. To the degree that these can be digitalized, there can be environmental as well as operational savings realized. When thinking of mobile-based technology, we are seeing (yet again) the development of e-readers. Some of these devices are also wireless devices that facilitate the distribution of printed material. The Kindle and Kindle II, by Amazon, the different Sony readers and a variety of Apple devices allow subscribers to download a wide variety of books, magazines, newspapers and other reading material over the network. This reduces the need for the distribution of the physical papers (and the consequent impact on the environment) and it also reduces the production/distribution costs for the newspaper companies.

A particularly burdensome aspect of the newspaper business is the physical distribution of the actual newspaper (Mitchell, 1996). It constitutes a significant part of the total cover price. Depending on the country and the number of ads, etc. the percent can range from approximately 25 to almost 45 percent of the cover cost of the newspaper (Wruck, 2006). An analysis of the production of the British paper *The Daily Mirror* shows that it produces 174 g of CO_2 per final newspaper sold. The manufacturing of paper is the most energy intensive portion of this process. This accounts for 70 percent of the CO_2 emission in the total process. The remaining 30 percent is associated with production, transportation/distribution and disposal.

The use of wireless devices and either the Internet or the mobile communications network address some of the environmental dimensions associated with the production of reading material. Mobile phone users are also exploring this possibility. For example, the state of California is moving to replace physical schoolbooks with electronic versions. This transition is not as revolutionary as it may sound. According to Wei, younger users and those who have an instrumental relationship to their mobile phones already view it as a news-gathering device (Wei, 2008). Indeed, the Internet is challenging the nature of traditional journalism. The ability to freely compose news stories from disparate sources while not needing to pay for the production of the material has resulted in a crisis for traditional newspapers. Thus, the digitalization of news has the potential to make a positive contribution in environmental terms, but this development is also opening the way for media production that undermines the traditional news gathering process.

Mobile Communications in Monitoring Climate Change and Disaster Response

One final area where mobile communications can make a contribution to the environment is in the area of monitoring climate change and in disaster response. In this case the devices are not mobile in the traditional sense, rather they are "stand alone" devices that can be easily placed in different critical locations and that communicate with a central hub using wireless protocols.

This type of "mobile" communication (device to device) can be used for example to monitor water flows in catchment areas where it is difficult to establish other types of monitoring. Similar systems can be applied to areas prone to bush fires. The systems can provide information on local moisture levels, humidity and wind. These elements could help in determining when conditions become critical. In the case of actual brush fires, these sensors would help in the development of a timely response.

If a natural disaster does indeed develop, mobile communication (now in the traditional sense of mobile phones etc.) has become an integral part of helping the victims find shelter during the disaster and in helping them contact loved one's after the disaster has passed (Palen et al., 2007).

Reducing the Carbon Footprint of Mobiles

Power, Recharging, and Batteries

One of the negative environmental impacts of mobile phones and other mobile devices need is the need for charging with electricity in order to function. The fact mobile phones have larger and more demanding displays and the fact that they also include radio senders/receivers for GPS, Bluetooth, and Wi-Fi in addition to the traditional mobile system means that the demands for electricity are increasing in the devices.

Mobile phones, as with other battery powered electronic devices are recharged on the mains in the developed world. In addition, if we leave our however, the chargers in the socket, they continue to draw a certain amount of so-called vampire or phantom power when the charger is in the socket and the device is either charged or absent from the charger.

The draw on the system is relatively small for each device, but the fact that there are many small contributions results in a relatively large demand on the power net. According to work done by Alan Meier at Lawrence Berkeley Labs, a mobile phone draws over 3.6 watts while charging and approximately 2.2 when charged but still connected to the charger. This is a relatively moderate amount compared to other appliances. Computers with their larger screens draw many times more than this as do set-top boxes, TVs, etc. It has been estimated that in the US (specifically California) that the average "low power" mode for different appliances was about 980 kilowatt-hours per year. This is about 13 percent of all electrical use (Redelmeier & Tibshirani, 1997).

In the case of mobile phones, there has been work at Nokia to develop chargers that do the work of charging the phone and then they turn themselves off, thereby reducing the "phantom" charging of the device. In the case of the Nokia charger, the user inserts the charger into a power socket, attaches the phone and then pushes a button to start the charging. When the phone is charged, it communicates with the charger and turns it off.

In the developed world, there is general access to stable power systems. In the developing world the situation is different. Indeed, there are many people living in developing countries that have limited access to electrical power. For those who do, the power is often unstable. In these cases, there are innovative services that are used to charge the batteries. Chipchase and Tulusan describe the development of local services where people can charge their phones (2007).

An obvious solution to this situation is the development of solar powered handsets. In an earlier phase of mobile phone diffusion in Bangladesh, solar power was used. In the Grameen "Village Phone" program, local "telephone ladies" received a small solar power device used to recharge the phone (Richardson et al., 2000). The diffusion of phones has meant that not just one person in the village has a phone and thus, there is an increased need for power. Sharp, Samsung and other producers have worked on the development of handsets that have a solar cell integrated onto the phone. If the talk time and the use of texting is somewhat moderate, these approaches are viable. Another approach is to use kinetic chargers that function on somewhat the same principle as self-winding watches such as that developed by M2E power. This said, these developments have not resulted in an appreciable number of handsets.

Another dimension of this issue is the development of mobile phones that are "greener" from the start. The main focus here might be on battery technology. Alternative power sources (perhaps fuel cells) would reduce the potentially toxic portions of mobiles and they would also have the potential of facilitating use developing countries.

Production, Use, and Recycling of Mobiles

Many of the items discussed above focus on the positive potentials for mobile communication vis-à-vis the environment. Ideas associated with reducing CO_2 emissions and providing for the more efficient distribution of resources are positive contributions.

The manufacture and use of a mobile phone also has an environmental impact. Following analysis of the work by Emmengger et al., it has been calculated that production of a mobile phone produces about 60 kg of CO_2 (Emmenegger et al., 2006). (Calculations are by the anonymous author of the "Fat knowledge" blog (Fat Knowledge, 2007).)

A computer consumes about 4.6 times as much energy to produce. Thus the 1.019 billion phones produced in 2006 resulted in about 60 million metric tons of CO_2 emissions. The impact of manufacturing a phone and using it for one year (including charging of the battery, running the base stations and switching equipment and administration of the system) is about 112 kg of CO_2. This assumes replacement of the phone after that one-year period. This is about 0.5 percent of the total CO_2 contribution of an average American (though it would be possible to substitute recycled materials in the production of the mobile phones).

Eventually the phone is retired. This may be the result of a contract period with a mobile provider ending, the phone may have quit functioning or there may simply be a new phone that has excited the passions of the user. If it is still functioning, the older phone might find new life as a replacement phone for another person or it might simply be put into a drawer along with other mobile phones that have lost favor (Wallis et al., 2009).

There are environmentally acceptable methods of dealing with mobile phones. On some occasions the phone, is refurbished and resold in another country. This happens only about 5 percent of the time (Huang & Truong, 2008). The Belgian company Unicom and the US company ReCellular have specialized in collecting unused phones, refurbishing those that can be used again and smelting those that are no longer serviceable (Mooallem, 2008).

The other alternative is that the phones are reclaimed for the valuable metals that they contain. If this is done responsibly, the toxic metals are also handled appropriately. In 2008 ReCellular reported collecting 5.5 million phones. This is indeed a large number, but it represents only a small portion of the more than 1 billion phones that are sold each year. In the processing of handsets that were no longer serviceable, ReCellular reclaimed 9500 kilos of copper, 432 kilos of silver and 43 kilos of gold. In addition, it reclaimed and processed the lead, arsenic, antimony, beryllium, cadmium, and other materials that were also a part of the mobile phones.

In spite of these efforts, the exporting of electronic waste is a serious issue. It is estimated that the US alone exports over 1.5 million tons of computers, TVs, VCRs, monitors, cell phones, and other equipment. Other parts of the developed world also contribute to this. In some cases inoperable equipment is exported under the guise of international aid. In other cases it is simple set overseas as junk. In either case, it has become an environmental challenge for the developing world. In developing countries, in order to salvage the reclaimable metals, this waste is burned in open fires or bathed in acids (Carroll, 2008). The resulting waste is obviously toxic and is a major hazard for those who live nearby.

Mobile Base Stations

Most people can relate to basic mobile services such as voice and SMS, and to the mobile equipment itself. The mobile network infrastructure is unfamiliar to most people. However, the infrastructure required to provide high quality voice and data traffic capabilities to end-users is very complex, costly and even has a significant carbon footprint at a worldwide scale. Indeed, when thinking about power consumption in a mobile telephone network, the handset uses only a minor portion of the total power budget. In addition to the core communication network, a series of base stations are placed in strategic places to ensure good coverage. Base stations are connected to a Mobile Switching Centre (MSC) and the rest of the core network through high capacity fiber channels to ensure enough bandwidth.

The base station deployment strategy is based on many factors, among which service capacity and service range are most important. Crowded public spots put additional constraints to the base stations requirements with regards to coverage and throughput. These constraints are often translated in higher power usage by the base station. A typical 2G/3G base station may consume up to 3 kW of power for legacy systems. Advancements in making more efficient Power Amplifiers (PAs) for base stations have reduced power consumption to less than half of that for the legacy systems. Many of the largest base station producers are presenting green solutions requiring less than 1 kW of power to operate. In addition, active on/off switching control based on traffic load reduced power consumption during idle mode. Such "green" features enables alternative energy sources such as wind and solar power generators to yield enough power to either power the base stations entirely, or for a long enough period to replace the environmentally unfriendly diesel generators as backup power suppliers.

However, to replace the equipment in all the base stations is a significant investment. Thus, while there is a great potential when building new base stations or in totally renovating outdated stations, there is a different economic calculation associated with upgrading active base stations.

The mobile communication infrastructure accounts for over 80 percent of the power used by the telecommunications industry. Speaking in terms of emissions, mobile base stations alone produce around 22 million tons of CO_2 annually.[1] A base station installed in areas where power supply is highly unreliable, or entirely off-grid, are often powered by diesel generators that annually consume up to 20,000 liters of diesel,[2] equivalent in excess of 50 kilotons of CO_2.

In short, the consumption of mobile phones is minor when compared to the consumption of the base station. If we use a rough estimate, there are about 2500 mobile phones "attached" to a standard mobile mast in a country such as Bangladesh. If we compute the consumption of the phones in comparison to the consumption of the mast, there is a scale of about 1:10 in terms of energy consumption and consequently CO_2. Thus, the real effort for reducing the carbon footprint of mobile communication might be best reduced by working on the consumption of the base stations. Recycled energy would also make good sense to the operators, particularly in the developing world, since base stations are often logistical problems. The reliance on diesel generators that require refilling and other logistical issues mean that economizing on the base stations would have a positive impact on the bottom line.

In more developed countries, there are often fewer phones tied to base stations. In Scandinavia, for example, there are more base stations since there is customer pressure, and licensing demands to have coverage along highways and in areas that are not necessarily tightly populated. This would simply increase the demand for efficient base stations. In such areas, however, solar energy is not as viable. Thus alternative forms of energy (wind, small scale hydro, etc) might be applicable.

Conclusion and Policy Suggestions

There are a series of areas where the application of mobile communication can potentially help to reduce the use of fossil fuels and thus reduce CO_2. In addition, there are a collection of issues that need to be confronted by mobile users in the pursuit of more environmentally friendly use. The specific issues include:
- Investigate the development of "packet based" traffic systems that allow for optimal traffic flow.

- Investigate the impact of micro-coordination on the generation or savings of transport in families.
- Mandate smart chargers
- Continue to investigate the development of more efficient base stations.
- Work to use alternative power in base stations where applicable
- Refinement of cyber-junk recycling systems
- Study when and where mobile communication facilitates better market information in developing countries and encourage that type of use.
- Pursue the development of more efficient base-station equipment and the powering of base stations with renewable energy.

In some cases, further research and development is needed. In other cases, such as the use of smart chargers, the technology is in place and should be implemented. As noted above, mobile communication is pervasive in society. Thus it represents a potential resource in the fight to reduce CO_2 pollution. These potentials and the problems with them need to be better understood and pursued.

Notes

1   http://www.inhabitat.com/2008/08/19/m2e-kinetic-energy-cell-phone-charger/.
2   Green Mobile Networks & Base Stations. Strategies, Scenarios and Forecasts 2009-2014. Juniper research http://news.cnet.com/8301-11128_3-9912124-54.html.

References

Barth, Matthew, Boriboonsomsin, Kanok, & Vu, Alex. (2007). *Environmentally-Friendly Navigation.* Paper presented at the Intelligent Transportation Systems Conference, 2007, Seattle, Washington.

Carroll, Chris. (2008, January 2008). High-tech trash. *National Geographic, 213,* 64-81.

Chipchase, Jan, & Tulusan, Indri. (2007). Street Charging Services Kampala.

Emmenegger, Mireille, et al. (2006). Life Cycle Assessment of the Mobile Communication System UMTS Towards Eco-Efficient Systems. *International Journal of Life Cycle Assessment, 11*(4), 265.

Fat Knowledge. (2007). Carbon Footprint of a Mobile Phone. Retrieved 29 April 2009, from http://fatknowledge.blogspot.com/2007/01/carbon-footprint-of-mobile-phone.html.

Gripsrud, Mattis. (2009). Kommunikasjonens tvillingpar. *Norsk Medietidsskrift, 16*(1), 25-47.

Horvitz, Eric, et al. (2005). *Prediction, expectation, and surprise: Methods, designs, and study of a deployed traffic forecasting service.* Paper presented at the Twenty-First Conference on Uncertainty in Artificial Intelligence, Edinburgh, Scotland.

Hsiao, Ruey-Lin, Wu, Se-Hwa, & Hou, Sheng-Tsung. (2008). Sensitive cabbies: On-going sense-making within technology structuring. *Information and Organization, 18*(4), 251-279.

Huang, Elaine, M., & Truong, Khai N. (2008). Sustainable ours: Situated sustainability for mobile phones. *Interactions, 15*(2), 16-19.

ITU. (2007). Mobile cellular, subscribers per 100 people. Retrieved 6.6.08, from http://www.itu.int/ITU-D/ict/statistics/.

ITU. (2008). Market Information and Statistics. Retrieved 30 January 2008, from http://www.itu.int/ITU-D/ict/statistics/.

Jolley, Ainsley. (2006). *Transport technologies*. Melbourne: Centre for Strategic Economic Studies, Victoria University.

Ling, Rich. (2004). *The Mobile Connection: The cell phone's impact on society*. San Francisco: Morgan Kaufmann.

Ling, Rich, & Donner, Jonathan. (2009). *Mobile phones and mobile communications*. London: Polity.

Ling, Rich, & Haddon, Leslie. (2003). Mobile telephony, mobility and the coordination of everyday life. In James E. Katz & Leopoldina Fortunati (Eds.), *Machines that become us*.

Ling, Rich, & Yttri, Birgitte. (2002). Hyper-coordination via mobile phones in Norway. In James E. Katz & Mark Aakhus (Eds.), *Perpetual contact: Mobile communication, private talk, public performance* (pp. 139-169). Cambridge: Cambridge University Press.

Manning, Peter K. (1996). Information technology in the police context: The 'sailor' phone. *Information systems research, 7*(1), 52-62.

Mitchell, William J. (1996). *City of bits: space, place, and the infobahn*: MIT press.

Mooallem, Jon. (2008, January 13, 2008). The Afterlife of Cellphones. *The New York Times Magazine*, from http://www.nytimes.com/2008/01/13/magazine/13Cellphone-t.html?_r=1&pa.

Nadeem, Tamer, et al. (2004). Trafficview: Traffic data dissemination using car-to-car communication. *ACM SIGMOBILE Mobile Computing and Communications Review, 8*(3), 6-19.

Palen, Leysia, Hiltz, Starr Roxanne, & Liu, Sophia B. (2007). Online forums supporting grassroots participation in emergency preparedness and response. *Communications of the ACM, 50*(3), 54-58.

Rasmussen, Terje. (2003). Mobile medier og individualisering. In K. Lundby (Ed.), *Flyt og forførelse. Fortellinger om IKT* (pp. 224-249). Oslo: Gyldendal Akademisk.

Redelmeier, Donald A., & Tibshirani, Robert J. (1997). Association between cellular-telephone calls and motor vehicle collisions. *The New England Journal of Medicine, 336*(7), 453-458.

Richardson, Ron, Ramirez, Ricardo, & Haq, Moinul (2000). *Grameen Telecom's Village Phone Programme in Rural Bangladesh: a Multi-Media Case Study*. Guelph Ontario: TeleCommons Development Group(TDG)o. Document Number).

Schrank, David L., & Lomax, Tim J. (2007). *The 2007 urban mobility report*: Texas Transportation Institute, Texas A & M University.

Townsend, Anthony M. (2000). Life in the real time city—mobile telephones and urban metabolism. *Journal of Urban Technology, 7*, 85-104.

Wallis, Cara, Qiu, Jack L., & Ling, Rich. (2009). *Alternative Mobile Handsets and Grassroots Innovation in the Global South: A Critical Examination*. Paper presented at the Association of Internet Researchers.

Wei, Ran. (2008). Motivations for using the mobile phone for mass communications and entertainment. *Telematics and Informatics, 25*(1), 36-46.

White, Peter B., & White, Naomi Rosh. (2009). Mobile telephones and social networking: CB Radio and the creation of transitory social networks in the Australian Outback. In Rich Ling & Scott W. Campbell (Eds.), *Mobile Communication: Bringing us together or tearing us apart?* New Brunswick: Transaction.

Wruck, Pierre. (2006). *Newspaper distribution today*. Paper presented at the Managing Distribution.

Yang, Liuging, & Wang, Fei-Yue. (2007). Driving into intelligent spaces with pervasive communications. *IEEE Intelligent Systems, 22*(1), 12.

6

Mobile Phones' Role following China's 2008 Earthquake

Yun Xia

On May 12, 2008, an earthquake of 8.0 magnitude struck two-thirds of China from an epicenter in Wenchuan, Sichuan Province. According to the relief progress report from the State Council Information Office of the People's Republic of China (2008), the quake claimed about 70,000 lives. Close to 18,000 people were missing and more than 370,000 were injured. In the most damaged areas, all telecommunications were severely affected. Given the massive destruction of infrastructure in the quake, people turned to mobile telephony for communication. On May 13, about one-third of the counties in the most damaged areas restored mobile communication and by May 20, almost 90 percent of the counties in the most damaged areas had no problem with mobile communication (Yang, 2008).

The current study incorporates culture in a framework to examine mobile phone use in a natural disaster. It explores the mutual interaction between Chinese culture and Chinese mobile phone use that sought information and aimed to reduce uncertainty about families, relatives, and friends after the earthquake. In this study, the pattern of mobile phone use is constructed and the implications for crisis communication policy are discussed.

Communication Technologies and Crisis Communication

Although the study of disaster has nearly a 100-year history, the exploration of information and communication technologies during recovery after a disaster has only happened recently (Mark & Senaan, 2008). With increased access to the Internet and various mobile technologies, people expand the possibilities of information sharing that is found to be critical in crisis communication after disasters (Seeger, Vennette, Ulmer, &

Sellnow, 2002; Palen & Liu, 2007). Information sharing serves rescue efforts, the recovery process, and post-crisis sense-making. Information sharing also reduces uncertainty about friends, family, and neighbors and helps decision-making in crises (Palen, Vieweg, Liu, & Hughes, 2009; Procopio & Procopio, 2007).

In the literature, information sharing and uncertainty reduction have been the key concepts used in exploring the relationship between people's information-seeking efforts and use of media, including newspaper, television, and the Internet after disasters. In the study by Boyle et al. (2004) of the interrelationship between emotional responses and increased use of the media after the terrorist attacks on 9/11 in the U.S., negative emotional response was a strong predictor of efforts to share information and reduce the discomfort and uncertainty. Procopio and Procopio (2007) concluded that communication technologies played an important role in reducing uncertainty, activating social networks, maintaining social capitals, and supporting geographically based communities during the crisis from Hurricane Katrina in 2005. With the reduction of uncertainty as the driving factor, Internet users from areas most damaged in the crisis spent more time online and engaged in more information-seeking activities than people from less damaged areas. Palen and her colleagues (Palen, Vieweg, Liu, & Hughes, 2009) described the instance of collective intelligence in decentralized and highly distributed information sharing, generating, and gathering through computer-mediated communications among peers after the shooting on the campus of Virginia Polytechnic Institute and State University in 2007.

In the literature, rarely is mobile phone use studied in crisis communication after disasters except for recent investigations of mobile phone communication as a part of citizen journalism in different natural disasters (e.g., Robinson & Robinson, 2007). However, a mobile phone is proven to allow people to keep in steady contact with families and friends in their social networks (Katz & Aakhus, 2002; Ling, 2004). A mobile phone can help people coordinate everyday activities, including setting up times and places to meet, seeking information, and refining social activities (Campbell & Kelley, 2006; Ling, 2004). The use of the three-way calling function on the mobile phone can reach out to people in different cities and coordinate shared activities (Campbell & Russo, 2003). This type of mobile phone use is defined as "instrumental uses" (Campbell & Kelley, 2006). On the other hand, mobile phones have various expressive uses that allow people to reach out to each other for social and emotional support. Expressive uses can take the form of chattering, gossiping, gabbing, discussing personal matters, and communications that seem meaningless in content but function to stay in touch (Campbell & Kelley, 2008; Johnsen, 2003; Licoppe, 2003). This leads to the following research question about Chinese use of mobile phone after the quake:

RQ1: How did Chinese use mobile phones to share information and reduce uncertainty in the crisis after the earthquake in Sichuan, China?

Cultural and Social Construction of Mobile Phone Use

Social constructionism advocates that social interaction shapes social behavioral patterns regarding technology use (Fulk & Boyd, 1991; MacKenzie & Wajcman, 1985). Through negotiations among various groups of people, not only are the meanings in people's use of technology constructed but they also are redefined in different social contexts (Leonardi, 2003). The key in the social construction approach of communication technology is social interaction that constructs the uses and effects of communication technology. In addition, individual's attitudes and behaviors tend to converge with others in the same social context because of shared symbolic systems and interpretive schemes. Thus, the use of communication technology is shaped by the social interaction it facilitates.

As the convergence of social, political, and economic contexts, culture shapes people's expression and drives their social interactions (Brock, 2005). Leonardi, Leonardi, and Hudson (2006) called for more attention to consider culture as an "integral part for the social construction process" (p. 223). Bell (2006) also reiterated the same idea when she studied mobile phone communication as "constellations of social and cultural practice" (p. 54). Goggin (2008) called for a "further, extended, and systematic and comprehensive studies" of mobile phone use in different national cultural contexts (p. 358).

There are a few studies that have examined the cultural meanings of Chinese use of mobile phones. In Wong's (2006) study of mobile phone use among senior residents in Hong Kong, mobile phones made it possible to create connections between generations even if there was little chance to get together. On the other hand, the maintained communication enabled elderly people and their children to have independence. In Law and Peng's (2006) study of the use of mobile phones among migrant workers in Southern China, mobile communication changed their consumption patterns, social relationships, and traditional values. On the one hand, traditional values were challenged because of the anonymous nature of the virtual relationships over mobile communication. On the other hand, mobile communication imposed constraints on migrant workers, reinforcing traditional village values and norms when they reconnected with their fellow villagers back home. In the three-year, multi-site ethnographic research, Bell (2006) explored how a mobile phone was used to maintain personal identities and social roles across seven different countries, including China. Her study revealed that the pre-existing patterns of social mobility outside of the home constructed through distinct cultural lenses shaped the adoption and use of mobile phones in these Asian countries. Xia's (2007) study about Chinese use of mobile phones analyzed the effect of Chinese cultural values and communication styles on different uses of mobile phones. The following research question incorporates Chinese culture in the evaluation of Chinese use of mobile phones in the crisis after the quake:

RQ2: How did Chinese culture shape Chinese use of mobile phones in the crisis after the earthquake in Sichuan, China?

Method

In this study, ethnography is used as the research method. Ethnography enables researchers to understand and describe a social and cultural scene from an insider's perspective. An ethnographic method includes fieldwork, selection of a place and a people to study, entry to the community of interest, non-participant observation and interviewing (Fetterman, 1998). Each process is described in the following design of the research.

Research Design

Fieldwork requires researchers to work with people for long periods of time in their natural settings. One of the researcher's friends works as a sales representative of China Mobile, one of the two largest mobile phone service providers in China. He was contacted and agreed to participate in the study. With his help, 300 invitations soliciting participation in the study were mailed to randomly selected subscribers of China Mobile in the city of Chengdu, the capital of Sichuan Province. Although the city is not in the most damaged area, it is only 40 miles from the quake epicenter. Eventually, 42 people responded and agreed to participate in the study. During a three-month period in the summer of 2009, the researcher visited the participants according to their addresses listed with China Mobile. Only 34 people finished the study with in-depth interviews.

In-Depth Interview

In order to keep the participants' meanings, the researcher chose non-directive interviews in which the participants talked in their own terms and controlled most of the interview conversations. In this way, the researcher hoped to find "unsolicited accounts" from the participants (Hammersley & Atkinson, 1995, p. 129). The researcher adopted Lofland and Lofland's (1984) idea about interviews as "guided conversations" (p. 85). Thus, there were no tightly structured questions. The researcher listed two categories that the researcher would talk about with each participant in the interviews. One was about the participants' use of mobile phones to share information and reduce uncertainty in the crisis after the quake. The other was about the relationship between Chinese culture and their use of mobile phones. Finally, all participants were invited for an in-depth interview. Each interview lasted from 30 to 40 minutes. All interviews were taped and transcribed. The interviews were conducted in Chinese and then translated into English.

Data Analysis

Hammersley and Atkinson's (1995) process for analyzing qualitative data was used. The first step was careful reading of interview transcripts. During the reading, special attention was paid to references to the participants' use of mobile phones in the crisis after the quake and how Chinese culture affected their use of mobile phones. Segments of the interview transcripts were reassembled and regrouped according to identified patterns (Fetterman, 1998). The analytical categories that were central to the study were generated. Examples and stories were used to illustrate analytical categories and patterns.

Findings

Three patterns surfaced from the data analysis: being difficult to get through; connection with family, relatives, and friends; and unique uses of mobile phones. Each pattern was also illustrated by different themes.

Being Difficult to Get Through

The reaction to the question about how people used mobile phones after the quake was "almost can't get through." Only a handful of the participants managed to call their families after the quake. Most people could not get through until 6 p.m., almost 4 hours after the quake.

First, the communication within Chengdu was down because of the dramatic increase of usage. When everyone tried to make a call in the crisis, the communication was severely congested and service disrupted. One participant, a student in the Sichuan Normal University, described his mobile phone use after the quake: "Everyone rushed out of the building. Boy, the building looks like a leaf in the wind swaying like hell. Man, while running, everyone took out the phone and tried to make a call."

Problems with mobile phones were at their worst half an hour after the quake. Some of the participants got through right after the quake. However, after half an hour communication became impossible. It took some time after the quake for mobile phone use to reach its peak, particularly for the communication from other areas of China to Chengdu. After the quake, communication from Guangdong Province to Sichuan Province increased by 1,000 times because there were millions of Sichuan migrant workers in Guangdong province. From Beijing to Chengdu, the increase was about 80 times. The average increase of mobile communication to Chengdu from other areas across the country was about 20 times (Yang, 2008). Even within Chengdu, use of mobile phones increased by 10 times. This increase is exceptionally high compared with the increased use of mobile phones in other crises. For example, in the crisis from the London bombings in 2005, calls rose by 67 percent and text messages increased by 20 percent (Gordon, 2007). This kind of increase was high enough to jam the mobile networks in the London area.

While the problem in Chengdu was increased traffic, damaged facilities in the most damaged areas made mobile communication impossible. After the quake, communication to the most damaged areas did not resume to normal until two or three days later. Mobile phone facilities include cellular towers, antennas, back-up batteries, cables, and power that make cellular towers work. After the quake, the most common cause for tower failure was damaged cables connecting towers. The second cause was power outage. By the morning of May 13, about 3,000 of China Mobile cellular towers had service failure in the most damaged areas (Yang, 2008). China Unicom had about 200 towers with service failure in those areas. One of the challenges in the post-quake relief work was to restore the damaged mobile phone facilities and the services of cellular towers in the most damaged areas.

Connection with Family, Relatives, and Friends

Reaching out to one's immediate family was the most important thing after the quake. In addition, contact with relatives and friends was also critical in the communication after the quake. Chinese cultural values about familial affinity drove people to contact their immediate family through mobile phones after the quake. In addition, the unique meaning of friendship in Chinese collectivism leads people to believe that close friends are no different from family.

During the quake, people rushed out of buildings, homes, schools, and stores. They went to open areas, such as town squares, fields, riverbanks, and stadiums. Many people had to sleep on streets even weeks after the quake because of aftershocks. Landline phones could not fulfill the communication task in this natural disaster. Mobile phones were the only communication medium that people could use to reach out to their families.

In addition to communication with family, the participants felt that they had to contact their relatives and friends. Interestingly, there was a priority list for contact after the quake. It was a list of friendship and familial affinity. On the list, family members, such as wives, husbands, children, and grandparents, were on the top. Close relatives, such as uncles, aunts, cousins, and nephews, were next. Close friends with frequent communications were the third on the list. All the other relatives and friends who were not so close were last.

The methods to make contact through mobile phones also varied according to the priority status that illustrated the negotiation process about technology use in the Chinese cultural context. For the immediate family members with the highest priority, voice calls through mobile phones were "a must." Voice calls are synchronous communication with instant feedback. People can find out about their families immediately, which fulfills the emergency need to share information and reduce uncertainty in the crisis after the quake. Meanwhile, voice calls are personal with their unique intonation, voice inflection, and use of natural language. In the literature, researchers have placed mobile phones between face-to-face interaction and landline phones in terms of richness of media communication (e.g., Campbell & Kelley, 2008). The media richness supports the participants' emotional needs in their personal, intimate communication with family members.

For relatives who were not the immediate family members, voice calls or text messaging were the two options for the participants. When it was difficult to get through, the participants switched to text messaging for sharing information about whether they were safe with their relatives. For the participants, relatives, particularly close relatives as uncles, aunts, and cousins, are still family members. By tracing the family tree, people value the concept that relatives are from the same family in China.

Thus, it was also urgent to make contact with one's close relatives after the quake. However, communication with them could be informational, not too personal.

The need to contact close friends was complicated by the participants' different understandings of friendship. Some participants insisted that close friends should be part of one's family. This understanding matches the importance of friendship in one's social networks in China. For this group of participants, immediate and personal communications with their close friends either through voice calls or text messaging was necessary. However, some other participants did not see the urgency to contact close friends. One participant said: "I guess I am a little selfish. I waited until the service got better to call my friends, even my close friends. Oh, for me, nothing is more important than family. What's the rush? Friends can be on the waiting list, especially when we cannot get through easily."

Most participants chose text messaging to contact their relatives and friends who were not close. In addition, they usually waited two or three days after the quake when the service became better. Text messaging was informational and meant to update friends and relatives about how they were doing.

Unique Mobile Phone Uses

After the quake, the driving force of information sharing and reducing uncertainty in a crisis, particularly uncertainty about family, relatives, and friends, led people to use mobile phones in ways that were not normal in everyday life.

First, in the case of service failure, people tried to track the signal of mobile phones to locate the whereabouts of their family. The researcher interviewed a lady from Helan province who had a business trip to Chengdu. Her story illustrated this unique way to reduce uncertainty about her parents after the quake. She was on the flight to Chengdu when the quake happened. Her parents had traveled to the Jiuzhaiguo Valley, a world-class tourism site that is 300 miles away from Chengdu, before the quake. She had lost contact with her parents. With help from the mobile service provider in her hometown, she got the code for the cellular tower from which her parents had their last contact with her. On her flight, another passenger knew people working for China Mobile in Chengdu. This passenger helped her retrieve the code of the location of the cellular tower. It turned out that the location was not in the most damaged area. The information made her relieved that her parents were probably safe.

An important mobile phone use was to turn them into mass media platforms in the natural crisis. Mobile phones are usually studied as an interpersonal communication medium (Campbell & Russo, 2003; Johnsen, 2003). In the communication after the natural crisis, the function of mobile phones was expanded in the juncture of social, technological, and cultural necessity.

The first use of mobile phones as a mass medium is the radio function on a phone. Participants talked about their use of radio on their mobile phones for the best information sources, such as quake updates, injury reports, relief progress, and location of lost families.

The relief workers also passed around radios for free in different communities in Chengdu after the quake. Radios are cheap, highly portable, and easy to use, with robust service when people are away from their homes on the streets and in stadiums, plazas, and other open areas for days and even weeks. As documented in the studies of crisis communication in the literature, radio broadcasters were found to abandon their established procedures and programs in favor of more immediate and responsive methods to disseminate information (Sellnow, Seeger, & Ulmer, 2002; Procopia & Procopio, 2007). After the quake, ratio stations in Chengdu combined for one collaborative broadcast, providing 24-hour coverage about quake damage and relief work progress. Radio stations also offered call-in shows that allowed the potential reunions of families and friends. Relief workers also used ratio stations to announce the need for relief manpower and facilities in specific locations.

Group text messaging turned mobile phones into a mass medium for the crisis communication. After the quake, all participants received text messages from their mobile phone service providers, either China Mobile or China Unicom, asking them to avoid using mobile phones intensively after the quake in order to save the limited mobile telephony resources for the relief work. One participant received a text message that read more like an advertisement: "Imagine someone underneath the quake debris who has a mobile phone on hand and wants to call 110 [emergency call number in China]. But the service is down because you, I, and he use mobile phones too much. Please cut off unnecessary calls and you will save lives."

The mobile phone service providers also broadcasted text messages with guidelines for phone use after the quake. One participant kept the message on his mobile phone: "First, use as little as possible. Second, do not redial if it does not get through. Third, speak as little as possible once it gets through. Fourth, no chitchatting, just messages about safety."

Another use of text message broadcasting was from schools, daycare centers, and kindergartens to parents. When the quake happened, it was about 2:30 p.m. Parents were desperate to reach out to their daughters and sons. In order to fulfill the information sharing need, text messages were sent massively to parents. One participant summarized the message from his son's middle school minutes after the quake: "The school did not get hit by the quake. Your child is safe. All students are on the school's track field. All classes are canceled. Parents should come in and pick up all their children immediately after reading this message."

Discussion

There are three patterns in Chinese use of mobile phones in the crisis after the quake: being difficult to get through; connection with family, relatives, and friends; and unique mobile phone uses. The findings answer the two research questions about how Chinese used mobile phones after the quake and how Chinese culture shaped the uses.

As indicated in the literature (e.g., Chu & Tang, 2006), Chinese use of mobile phones can be understood better when the cultural specifications are considered as the context. In the first and second patterns that underline people's urgent need to contact families, relatives, and friends after the quake, the cultural values of familial affinity and friendship in Chinese collectivism play an important role. Collectivism emphasizes solidarity and integration with others and prioritizes the needs of the group over the needs of the self. Being collectivistic, Chinese culture emphasizes in-group belonging, personal interdependence, and social harmony (Lu, 1997). Chinese culture demands that individuals fit into the group and prioritizes the common good of the whole over individual wishes (Yang, 1994). All these values dictate the unique importance of family in Chinese culture. In addition, Chinese understanding of friendship in the personal relationship networks summarizes the reliance on harmonious personal relationships in Chinese culture. Connections and contacts with friends are the key to cultivate and maintain personal relationship networks.

In addition, as the underlying logic regarding the use of personal communication technologies, perpetual contact (Katz & Aakhus, 2002) is the notion of pure communication or the ideal of sharing one's mind constantly with another. Perpetual logic is a socio-logic that "underwrites how we judge, invent, and use communication technologies" (p. 307). Personal communication technologies including mobile phones offer people the means for perpetual contact and people use these technologies in a coherent way across different countries.

The logic of perpetual contact finds a perfect fit in Chinese cultural values about families, relatives, and friends. This fit drove millions of people to take out their mobile phones and try to make contact with a long list of people immediately after the quake, which created the explosive increase of mobile phone use. This fit also explains why people kept redialing when they could not get through, which made the congested communication worse during a natural crisis.

After the quake, it is critical for the government to construct an alternative emergency communication network. The explosive increase of mobile phone use after the quake that congested the mobile communication has proven the vulnerability of the public telecommunication network in natural disasters. The development of the alternative mobile communication systems has been widely discussed as a part of the important telecommunication policy change in China (Li, Ma, & Hu, 2009; Zheng & Zheng, 2008; Zi, 2009). One of them can be the further development of satellite communication networks that are robust in a crisis communication. Satellite communication was used in the quake relief work including coordination of communication reconstruction, live news coverage, and temporary communication vehicles. After the quake, the communication out of the epicenter town of Yingxiu in the city of Wenchua was cut off completely. After rescue workers managed to deliver 10 satellite phones to the township, the first phone call was made from the epicenter about 18 hours after the quake (Zi, 2009). However, there are only 5 communication satellites serving China while 8 satellites helped the Kobe quake relief in Japan in 1995 and 120 satellites were involved in the Hurricane Katrina relief work in the U.S. (Zi, 2009). One of the policies is for the government to speed up the construction of satellite communication systems. The key is to make sure the equipment of satellite communication terminals in the counties and villages where natural disasters happen more often as part of the undergoing telecommunication construction project, "no village left behind in communication," in China. While serving as the communication system among the rural and remote areas, satellite communication can meet the communication challenges during natural disasters. In the relief plan for natural disasters published by the Central Government of the People's Republic of China (2006), the role of satellite communication was clearly emphasized in the natural disaster warning, monitoring, and relief process.

In the third pattern, the unique mobile phone uses match the emergence of novel communication systems in a crisis that are often featured by various creative and flexible uses of media (Sellnow, Seeger, &

Ulmer, 2002). The participants' use of the radio function on their mobile phones proves that radio is one of the best relief communication media. Among the recommendations about what to do after a quake from the U.S. Federal Emergency Management Agency (FEMA, 2010), one of them is to use battery-operated radios for the latest emergency and relief information.

Chinese use of massive test messaging after the quake turned mobile phones into a mass medium. Bell's (2006) study revealed one of the cultural reasons why text messaging is very popular in China as well as other Asian countries. Chinese use text messaging to communicate all kinds of information outside formal channels. They send holiday greetings, share jokes, and communicate messages that cannot be talked about openly, such as sex or politics. Because of the ideographic nature of Chinese language, the limited space slots on a mobile phone are filled with Chinese characters, each of which is a word. Thus, Chinese can have "a far richer messaging experience" than non-character based languages, such as English, that fill spaces with letters and need multiple spaces for a word (p. 42). The use of massive text messaging is the direct result of the interaction between Chinese users' cultural and social needs and communication technology in the crisis communication after the quake.

The successful use of massive text messaging after the quake brings a controversial argument about text messaging in the policy discussion forum. The government does not always support the use of text messaging in crisis communication when there is a need to control information sharing. During the crisis after the SARS outbreak, the official policy of cover-up and unresponsiveness in the early stages resulted in a strict censorship of messages sent over mobile phones as well as the Internet. There were over 2800 facilities across China that monitored text-messaging traffic. If the topic was considered to be sensitive, the message was not transmitted (Gordon, 2007). However, through text messaging, people could bypass official control of vital media channels and share information about the location of outbreaks, development, prevention and possible cures. An alternative discourse was created when text messaging became an outlet for people's discontent and concern with the government's inattention and unresponsiveness to the epidemic in the early stages (Gordon, 2007). When information sharing is necessary for relief work in a crisis, text messaging has been discussed as one of the important relief communication methods in the telecommunication industry (Zheng & Zheng, 2008). One of the policies is to urge the mobile service providers to use text messaging informing their users about the crisis or emergency and encourage their users to switch to text messaging as the exclusive communication method during the crisis.

Future Research

This study had 34 participants. Although small samples are common for in-depth qualitative studies, the generalizability of the findings is unknown to other groups of people. Further examination with quantitative methods is needed with a large sample size. The relationship between Chinese culture and mobile phone use may not be generalized to other communication technologies, such as the Internet. More studies are needed to compare the uses of different communication technologies in a crisis since people rarely use one communication technology exclusively in the communication process.

In addition, the use of mobile phones as a mass media communication technology represents an important area for future research. As an interpersonal communication technology, the mobile phone is studied for micro-coordination and hyper-coordination of social activities (Ling & Yttri, 2002). In the current study, massive text messaging turned mobile phones into a mass media communication platform. Communication scholars might focus on the ways in which mobile phones function to disseminate news and information, particularly in a crisis communication process.

Another area for future research involves the comparison of different kinds of disasters. The nature of a crisis can shape the ways that people use communications. The exploration of communication processes in floods, wildfires, and hurricanes may help clarify the differences of mobile phone use in crises.

Conclusion

The current study includes culture in a framework to examine Chinese use of mobile phones in the crisis after the quake on May 12, 2008 in Sichuan, China. From the ethnographic fieldwork, three patterns are constructed to tell different stories about Chinese use of mobile phones to share information and reduce uncertainty while reaching out to families, relatives, and friends. In addition, in the context of cultural specifications, Chinese use of mobile phones in the crisis is understood better. The juncture between communication technologies, such as mobile phones, and culture leads to the distinct practices that illustrate people's construction process of a particular communication technology use in a society.

References

Bell, Genevieve. (2006). The age of the thumb: A cultural reading of mobile technologies from Asia. *Knowledge, Technology, & Policy, 19*(2), 41-57.

Boyle, Michael P., Schmierbach, Mike, Armstrong, Cory L., McLeod, Douglas M., Shah, Dhavan V., & Pan, Zhongdang. (2004). Information seeking and emotional reactions to the September 11 terrorist attacks. *Journalism and Mass Communication Quarterly, 81*, 155-167.

Brock, André. (2005). "A belief in humanity is a belief in colored men:" Using culture to span the digital divide. *Journal of Computer-Mediated Communication, 11*(1), article 17. Retrieved from http://jcmc.indiana.edu/vol11/issue1/brock.html.

Campbell, Scott. W., & Kelley, Michael J. (2006). Mobile phone use in AA networks: An exploratory study. *Journal of Applied Communication Research, 34*(2), 191-208.

Campbell, Scott W., & Kelley, Michael J. (2008). Mobile phone use among alcoholics anonymous members: New sites for recovery. *New Media and Society, 10*(6), 915-933.

Campbell, Scott W., & Russo, Tracy C. (2003). The social construction of mobile telephony: An application of the social influence model to perceptions and uses of mobile phones within personal communication networks. *Communication Monographs, 70*, 317-334.

Chu, Wai-chi, & Yang, Shanhua. (2006). Mobile phones and new migrant workers in a South China village: An initial analysis of the interplay between the "social" and the "technological.' In Pui-lam Law, Leopoldina Fortunati, & Shanhua Yang (Eds.) *New technologies in global society* (pp. 221-245). Singapore: World Scientific.

Fetterman, David M. (1998). *Ethnography: Step by step.* Thousand Oaks, California: Sage.

FEMA (2010). *What to do after an earthquake.* Retrieved from http://www.fema.gov/hazard/earthquake/eq_after.shtm.

Fulk, Janet & Boyd, Brian. (1991). Emerging theories of communication in organizations. *Journal of management, 17*, 407-447.

Goggin, Gerard. (2006). Cultural studies of mobile communication. In James E. Katz & Manuel Castells (Eds.) *Handbook of mobile communication studies* (pp. 353-366). Cambridge, MA: The MIT Press.

Gordon, Janey. (2007). The mobile phone and the public sphere: Mobile phone usage in three critical situations. *Convergence: The International Journal of Research into New Media Technologies, 13*(3): 307-319.

Hammersley, Martyn, & Atkinson, Paul. (1995). *Ethnography: Principles in practices.* New York City, New York: Routledge.

Johnsen, Truls E. (2003). The social context of the mobile phone use of Norwegian teens.

In James E. Katz (Ed.), *Machines that become us: The social context of communication technology* (pp. 161-70). New Brunswick, NJ: Transaction Publishers.

Katz, James E., & Aakhus, Mark A. (2002). Conclusion: Making meaning of mobiles—a theory of Apparatgeist. In James E. Katz & Mark A. Aakhus (Eds.), *Perpetual contact: Mobile communication, private talk, public performance* (pp. 301-320). Cambridge, UK: Cambridge University Press.

Law, Pui-lam & Peng, Yinni. (2006). The use of mobile phones among migrant workers in southern China. In Pui-lam Law, Leopoldina Fortunati, & Shanhua Yang (Eds.), *New technologies in global society* (pp. 245-258). Singapore: World Scientific.

Leonardi, Paul M. (2003). Problematizing "New Media": Culturally based perceptions of cell phones, computers, and the Internet among United Sates Latinos. *Critical Studies in Media Communication, 20*, 160-179.

Leonardi, Paul M., Leonardi, Marianne E., & Hudson, Elizabeth. (2006). Culture, organization, and contradiction in the social construction of technology: Adoption and use of the cell phone across three cultures. In Anandam P. Kavoori & Noah Arceneaux (Eds.), *The cell phone reader: Essays in social transformation* (pp. 205-225). New York: Peter Lang Publishers.

Li, Guangxia, Ma, Chong, & Hu, Jing. (2009). *Wei xing tong xin zai tu fa shi jian zhong de ying yong* [Satellite communication technology applications in emergencies]. *Shu zi tong xin shi jie* [*Digital Communication World*], 5, 36-39.

Licoppe, Christian. (2003) Two modes of maintaining interpersonal relations through telephone: From the domestic to the mobile phone. In James E. Katz (Ed.), *Machines that become us: The social context of communication technology* (pp. 171-86). New Brunswick, NJ: Transaction Publishers.

Ling, Richard S. (2004). *The mobile connection: The cell phone's impact on society.* San Francisco, CA: Morgan Kaufman Publishers.

Ling, Richard S. & Yttri, Birgitte. (2002). Hyper-coordination via mobile phones in Norway. In James E. Katz& Mark A. Aakhus (Eds.), *Perpetual contact: Mobile communication, private talk, public performance* (pp. 139-69). Cambridge: Cambridge University Press.

Lofland, John & Lofland, Lyn H. (1984). *Analyzing social settings: A guide to qualitative observation and analysis.* Belmont, CA: Wadsworth.

Lu, Shuming. (1997). Culture and compliance gaining in the classroom: A preliminary investigation of Chinese college teachers' use of behavior alteration techniques. *Communication Education, 46,* 10-28.

MacKenzie, Donald, & Wajcman, Judy. (Eds.). (1985). *The social shaping of technology: How the refrigerator got its hum.* Milton Keynes, UK: Open University Press.

Palen, Leysia & Liu, Sophia B. (2007) Citizen communications in disaster: Anticipating a future of ICT-supported public participation. In *Proceedings of the SIGCHI Conference on Human Factors in Computing Systems* (pp. 727-736). New York, NY: ACM Press.

Palen, Leysia, Vieweg, Sarah, Liu, Sophia B., & Hughes, Amanda L. (2009). Crisis in a networked world: Features of computer-mediated communication in the April 16, 2007 Virginia Tech Event. *Social Science Computing Review*, 27, 467-480.

Procopio, Claire H., & Procopio, Steven. T. (2007). Do you know what it means to miss New Orleans? Internet communication, geographic community, and social capital in crisis. *Journal of Applied Communication Research, 35* (1), 67-87.

Robinson, Wendy, & Robinson, David. (2007). Tsunami mobilizations: Considering the role of the mobile and digital communication devices, citizen journalism, and the mass media. In Anandam P. Kavoori & Noah Arceneaux (Eds.), *The Cell Phone Reader: Essays in Social Transformation* (pp. 85-103). New York, NY: Peter Lang Publishing Group.

Seeger, Matthew W., Vennette, Steven, Ulmer, Robert R., & Sellnow, Timothy L. (2002). Media use, information seeking, and reported needs in post crisis contexts. In Bradley S. Greenberg (Ed.), *Communication and terrorism: Public and media responses to 9/11* (pp. 53-63). Cresskill, NJ: Hampton Press.

Sellnow, Timothy L., Seeger, Matthew W., & Ulmer, Robert R. (2002). Chaos theory, informational needs, and natural disasters. *Journal of Applied Communication Research, 30,* 269-292.

State Council Information Office of the People's Republic of China. (2008). *September 25 relief progress report.* Retrieved from http://www.scio.gov.cn/zt2009/dzzqx/01/04/200905/t304560.htm.

The Central Government of the People's Republic of China. (2006). Guo jia zi ran zai hai jiu zhu ying ji yu an [*National emergency plans for natural disaster relief*]. Beijing, China. Retrieved from http://www.gov.cn/yjgl/2006-01/11/content_153952.htm.

Wong, William Wai-lim. (2006). Mobile phones, aged homes, and family relations in Hong Kong preliminary observations. In Pui-lam Law, Leopoldina Fortunati, & Shanhua Yang (Eds.), *New technologies in global society* (pp. 179-193). Singapore: World Scientific.

Xia, Yun. (2006). Cultural values, communication styles, and use of mobile communication in China. *China Media Research, 2*(1)*, 24-38.

Yang, Mayfair Mei-hui. (1994). *Gifts, favors, and banquets: The art of social relationships in China.* Ithaca, NY: Cornell University Press.

Yang, Yuliang. (2008). Ke ji zhi jiu kang zhen jiu zai wang [Construction of the earthquake relief network from technology]. *Ruan jian gong cheng shi [Software Engineers], 7,* 12-21.

Zheng, Dayong, & Zheng, Hongjian. (2008). *Di zhen gao su wo men sheng mo? Tong xing yie in dui tu fa shi jian de si kao.* [What the earthquake tells us: Telecommunication in emergences]. *Zhong guo dian xin yie [China Telecommunications], 91,* 19-21.

Zi, Xiao. (2009). Da zao tian di yi ti hua de wei xin tong xing xi tong. [Space and ground: Creating an integrated satellite communication system]. *Tai kong tan suo [Space Exploration], 9,* 16-19.

7

Social Networks and Policy Knowledge during the 2008 US Presidential Election

Scott W. Campbell and Nojin Kwak

It is now well established that mobile communication has become a primary resource for strengthening personal network ties. While these connections offer certain social capital benefits, scholars also question whether intensive use of the technology among these ties might lead to withdrawal from public aspects of social life (Campbell et al., 2010; Campbell & Kwak, 2010, in press; Gergen, 2008; Ling, 2008; Wilken, 2010). Gergen (2008) advances the notion of "monadic clusters" to characterize a cultural trend toward continual mobile-mediated connection in small enclaves of like-minded network ties. He argues that heavy use of mobile technology in these types of networks can foster insularity, hindering public dialogue and participation in the democratic process. There may be harmful effects on the policy environment when closed systems of circular affirmation hamper questioning, deliberation, and exposure to other sources of dialogue about social issues and their policy implications.

This chapter reports on an empirical study of how mobile communication with strong network ties is associated with knowledge of candidate policy positions during the 2008 US presidential election. As expected, mobile communication intersects with the characteristics of strong ties in distinctive ways to predict policy knowledge. Increased use of the technology in networks that are small and in those that are like-minded were both associated with less policy knowledge. In contrast, knowledge levels increased significantly with mobile-mediated discussion among politically diverse network ties. Collectively, the findings suggest both promise and peril for the policy environment, depending on levels of mobile phone use as well as characteristics of the networks in which it is used.

In 2005 the Center for Mobile Communication Studies held its second conference, culminating in the *Handbook for Mobile Communication Studies* (Katz, 2008). In a plenary session and chapter in the edited volume, Kenneth Gergen (2008) advanced theoretical propositions about how mobile communication is transforming civil society, with important—and not altogether rosy—consequences for the democratic process. This chapter examines Gergen's propositions to further our understanding of how mobile communication, as an emergent channel for social networking and political communication, may have helped and hindered policy knowledge during the 2008 US presidential election.

Essentially, Gergen argues that intensive use of mobile technology can lead to social network insularity with deleterious consequences for open deliberation and political participation. As we will discuss later, this hypothesis has been tested through recent research, lending some support to Gergen's concerns while also highlighting new opportunities of mobile technology (Campbell et al., 2010; Campbell & Kwak, in press). In addition to deliberation and participation, there is a third cornerstone of civil society that should also be considered—knowledge. Political knowledge is a crucial resource for the democratic process. It enhances the depth and breadth of debate, aids opinion formation, and helps inspire and guide political involvement. Therefore, we propose knowledge be included in discussions about the health of civil society and the potential dangers of network insularity. In this study, we are particularly interested in knowledge of candidate policy positions during the last national election. Among other forms of political knowledge, awareness of policy matters helps informed decision making and raises the quality of political discourse. Therefore, this chapter reports on a study examining how candidate policy knowledge might have been supported or restricted by mobile communication among network ties during the election.

Like others, Gergen (2008) recognizes the benefits of this new channel for voicing opinions and coordinating demonstrations. However, he also sees mobile technology as contributing to a structural shift in civil society where autonomous individuals are replaced by small enclaves of close ties. Gergen argues that continual contact through the mobile phone can lead to "monadic clusters," or small, insular networks of like-minded individuals. His central concern is that one is connected to their strong ties at the expense of political life. Gergen suggests that political detachment can occur when the dominant interests within these networks are with the immediate lives of the members themselves. He acknowledges that some may share an interest in political matters, but contends they will coalesce in their views through circular affirmation, thus hindering open dialogue and exposure to alternative perspectives.

Gergen's notion of the monadic cluster extends from research highlighting the prominent use of mobile technology for strengthening networks of strong social ties (e.g., Ling, 2004, 2008). In this context, mobile communication is a personalized resource that supports "selective sociality" (Matsuda, 2005). Indeed, there is evidence that such use of the technology can lead to "tele-cocoons" (Habuchi, 2005) of closely knit and like-minded networks, raising concerns about withdrawal from the public realm of social life (Campbell et al., 2010; Campbell & Kwak, 2010, in press; Ling, 2008; Wilken, 2010).

Elsewhere, we have reported findings that support the notion of monadic clusters, while at the same time tell a very different story about advantages of mobile communication among network ties (Campbell et al., 2010; Campbell & Kwak, in press). As Gergen anticipates, our previous work found that political involvement and open dialogue decline with intensive use of the mobile phone for discourse in small networks of like-minded individuals. However, participation and open dialogue both increase with mobile-mediated discourse in larger homogenous networks. Thus, mobile communication enhances positive as well as negative trends associated with social network characteristics. In other words, its capacity to help or hinder civil society is dependent not just on how the technology is used, but whom it is used with (Campbell et al., 2010; Campbell & Kwak, in press).

In formulating a hypothesis, we draw from the findings discussed above and the theory of monadic clusters to predict that intensive mobile communication among strong personal ties in smaller, more like-minded networks will be linked to less knowledge of candidate policy positions. Testing this hypothesis will involve looking at the interaction effects that network size and like-mindedness, when combined, have on the association between mobile communication and policy knowledge. Because mobile communication has the capacity to help as well as hinder civic life, we also consider scenarios in which it may support policy knowledge.

The direct relationship between mobile communication with strong ties and policy knowledge will serve as an initial baseline for examining trends. Research demonstrates discussion networks are a key ingredient to knowledgeable and active citizenship. Discourse leads to various forms of participation (Eveland & Hively, 2010; Kwak, Williams, Wang, &

Lee, 2005; McLeod, Scheufele, & Moy, 1999) and aids in the under-standing of political issues Kwak et al., 2005; Scheufele, Shanahan, & Kim, 2002). Considering these benefits of discussion networks, we anticipate that mobile communication with close ties will be positively linked to policy knowledge.

With regard to the interactive effect of size, previous studies show that members of larger discussion networks tend to be more politically active and knowledgeable (Campbell & Kwak, in press; Kwak et al., 2005; Mutz, 2002a, 2002b). Conceivably, individuals are better able to tap into the benefits of a large network of strong ties when connected through the mobile phone, which provides flexibility in when, where, and how often these individuals can be reached. Therefore, we anticipate mobile communication to be positively linked to policy knowledge for members of networks containing a large number of close ties.

As for like-mindedness, political theory has traditionally favored the assumption that a diverse discussion environment has a positive impact on civil society. As Arendt (1968) and Habermas (1989) contend, exposure to alternative views fosters interpersonal deliberation and reflection, which enhances one's ability to form opinions. Exposure to opposing perspectives is thought to promote greater awareness of and appreciation for alternative arguments, thereby providing individuals with a more robust and sympathetic framework for making decisions about political matters (Benhabib, 1992; Manin, 1987). Heterogeneous networks provide greater access to knowledge that equips individuals to be engaged and informed citizens (Scheufele et al., 2006). Extending this line of thinking to a mobile communication context, we anticipate higher levels of policy knowledge with increased mobile communication in networks containing a variety of political perspectives.

Survey of Policy Knowledge during the 2008 US Presidential Election

In order to test these hypotheses, we conducted a mail panel survey just prior to the 2008 presidential election yielding a nationally repre-sentative sample of 1,012 adults in the US. Among demographics and a host of other measures, the survey assessed how often individuals used the mobile phone for connecting with strong network ties, the size of these networks, and the extent to which the members are politically like-minded. The survey also contained a battery of questions testing respondents' knowledge of policies supported by candidates Barack Obama and John McCain.

Mobile communication with strong ties was measured with seven items asking how often in the last month respondents used the technology with these individuals to: (1) coordinate or make plans, (2) catch up or stay in touch, (3) place or receive a voice call, (4) exchange text/instant messages, (5) exchange emails, (6) exchange pictures, and (7) discuss politics and public affairs. Participants selected from an eight-point response option for each ranging from "never" to "several times a day," and the responses were formed into an additive index ($M = 21.28/56$, $SD = 11.39$, Cronbach's alpha $= .88$). A definition of strong network ties preceded this portion of the survey, characterizing them as individuals one feels very close to and comfortable with in discussing feelings and personal issues. For network size, participants were asked to write in the number of strong ties they have ($M = 8.55$, $SD = 6.19$). Like-mindedness was assessed using three items asking respondents how many of their strong ties (1) share the same share political views, (2) support the same presidential candidate, and (3) believe the same issues are important. Response options were on a five-point scale ranging from "none" to "all" and combined into an additive index ($M = 9.96/15$, $SD = 2.92$, Cronbach's alpha $= .86$). The items for policy knowledge asked respondents whether Barack Obama and John McCain support the following policies: (1) direct presidential diplomacy with Iran without pre-conditions, (2) offshore drilling as a key solution to the energy crisis, (3) privatized health savings account as part of health care reform, and (4) increase in income taxes for those earning over $250,000 annually. Correct answers received a value of 1, while incorrect responses and those indicating "don't know" were coded as zero. These values were then added for a composite score for policy knowledge ($M = 3.42/8$, $SD = 2.50$).

Using regression procedures, we examined the links between mobile communication and policy knowledge while accounting for the effects of social network characteristics by including them as interaction terms. In presenting the analysis, we will build our way up by first reporting on the direct relationship between mobile communication and policy knowledge, followed by the moderating effects of network size and like-mindedness individually, and then the three-way interaction effect of mobile communication combined with both of these characteristics together.

Findings and Implications

When looking at mobile communication without network characteristics, we found an unexpected result. In contrast to what we anticipated, increased mobile communication with strong ties did not contribute to policy knowledge for participants during the 2008 presidential election. In fact, it may have detracted from it. Table 7.1 shows a negative relationship between mobile phone use with strong ties and policy knowledge prior to accounting for network characteristics, although it is not technically significant ($p = .20$). This finding is unexpected because discussion networks typically contribute to, rather than detract from, understanding of political issues (Kwak et al., 2005; Scheufele et al., 2002). Furthermore, in our previous research, mobile-mediated discourse with close ties was positively linked to political involvement (Campbell & Kwak, in press). Although, in that study we focused on informational uses of the technology, whereas in this case we are examining all types of use with strong ties. Still, it is conceivable that mobile communication may have differential effects in different areas of political life.

Table 7.1 Predictors of Policy Knowledge

	β	t Value	Sig.
Control Variables (Block 1), $R^2 = .21$			
Age	−.00	−.07	.47
Gender (high: female)	−.11	−.38	.00
Education	.18	5.72	.00
Household income	.08	2.57	.01
Political interest	.32	10.10	.00
Newspaper news use	−.04	−1.15	.13
Television news use	−.01	−.44	.66
Strong Tie Variables (Block 2), Incremental $R^2 = .01$			
Mobile Use w/ Strong Ties	−.03	−.85	.20
Size	.04	1.31	.09
Like-minded	.10	3.20	.00
Two-Way Interactions (Block 3), Incremental $R^2 = .01$			
Mobile Use X Size	.03	1.12	.13
Mobile Use X Like-minded	−.06	−2.00	.02
Three-Way Interaction (Block 4), Incremental $R^2 = .00$			
Mobile X Size X Like-minded	−.01	−.386	.35

Note: Entries for control and strong tie variables are standardized final regression coefficients when the variables in the blocks are analyzed. Entries for the interaction variables are upon-entry Betas after controlling for prior blocks. All significant tests are one-tailed.

While this finding is noteworthy, it reveals little about the implications of monadic clusters since it only accounts for one of their key characteristics, i.e., frequent mobile communication. Therefore, we included network size and like-mindedness to better understand how they interact with mobile communication in distinctive ways to predict individuals' levels of policy knowledge. These tests were performed in two stages. First we examined the interactive effects of size and like-mindedness individually on the links between mobile communication and policy knowledge. After these two-way interactions, we ran a test of the three-way interaction between size, like-mindedness, and mobile communication together.

With regard to the interactive effect of mobile communication X size, Figure 7.1 shows the trend for policy knowledge is nearly flat with increased mobile contact with a large number of strong network ties, but declines notably with increased mobile communication in smaller networks. This pattern resonates with Gergen's warning that intensive mobile contact in small networks can cut individuals off from the democratic process. In this case, it is possible that "perpetual contact" (Katz & Aakhus, 2002) with a small handful of network ties can limit exposure to information about politics and public affairs with negative consequences for policy knowledge. One might expect (as we did) that an increase in network size would reverse this trend because it provides more opportunities for discussing the election or other topics that might lead to increased knowledge. However, this is not the case. Overall, this finding highlights a potential danger of mobile communication in smaller networks, which is a salient concern considering mobile communication is on the rise, while the average size of core discussion networks has been trending downward in the US (McPherson, Smith-Lovin, & Brashears, 2006). While we consider the interactive effect of small network size meaningful for the purposes of this discussion, conclusions must be cautioned by a marginal significance level of .13.

Figure 7.1 Predicting Policy Knowledge with Mobil Communication X Network Size

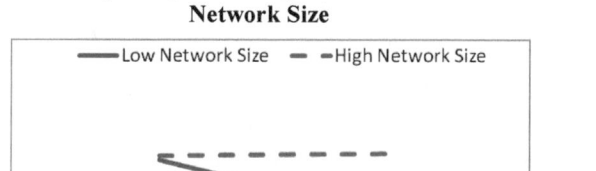

The finding for the interactive effect of like-mindedness shows a downward trend for policy knowledge with increased mobile phone use among network ties who share the same views, and a moderately positive trend for use with strong ties who have differing political views (see Figure 7.2). This set of findings suggests that mobile communication with close ties may either support or detract from policy knowledge, depending on how like-minded those ties are. In this case, there is nothing advantageous about network diversity per se. In fact, those with like-minded strong ties have higher levels of policy knowledge than those with more politically diverse ties, in cases of low mobile phone use. However, as mobile communication increases, so too does policy knowledge for those in more diverse networks, while at the same time declining for those in like-minded networks. Thus, there is a crossing pattern in Figure 7.2 that suggests both promise and peril of the technology by having a synergistic effect with both positive and negative trends associated with network characteristics. The disadvantages of like-mindedness and advantages of political diversity are only apparent when increased mobile communication is thrown into the mix.

Figure 7.2 Predicting Policy Knowledge with Mobile Communication X Like-minded

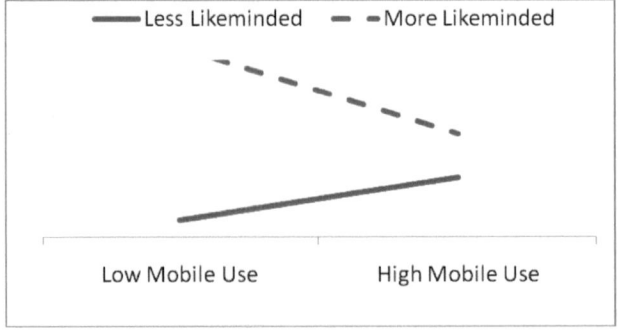

The two-way interaction effects of size and like-mindedness show how social network characteristics have an important influence on the links between mobile communication and policy knowledge. When network size and diversity are left out of the picture, mobile contact with strong ties did not lead to a greater awareness of candidate positions during the election. Because this diverges from the typically positive outcomes of discussion networks, it might be tempting to conclude that there is something distinctive about this medium that reduces those benefits.

These types of arguments have been made about computer-mediated communication as well (e.g., Daft & Lengel, 1986). However, the interactive effects show that it is hasty to make sweeping claims about mobile communication without considering how it plays off of and feeds into social network characteristics. This point is clearly illustrated by the crossing patterns where trends for policy knowledge are reversed with variations in network size and like-mindedness. These findings can be understood through the lens of social construction. That is, perceptions, uses, and implications of electronic social media are substantially shaped by one's network of close ties (Campbell & Russo, 2003; Fulk, Schmitz, & Ryu, 1995; Fulk, Schmitz, & Steinfield, 1990). Accordingly, whether mobile communication technology helps or hinders civil society and the policy environment is determined by the social context in which it used and the social contacts it is used with.

So far we have examined the moderating effects of each social network characteristic separately, meaning we have not yet fully captured the concept of a monadic cluster, which combines these dimensions. As noted, monadic clusters are small enclaves of like-minded individuals who heavily rely on mobile communication, with harmful effects on political participation and deliberation (Gergen, 2008). Because this proposition has received some support in the research (Campbell et al., 2010; Campbell & Kwak, in press), we anticipated a negative relationship between mobile communication with strong ties and policy knowledge in networks that are both small *and* like-minded. However, findings for the three-way interaction effect of mobile communication, size, and like-mindedness together were not statistically significant ($p = .35$). In other words, the combination of all three key aspects of the monadic cluster was not associated in a meaningful way with levels of policy knowledge. We consider this to be good news for policy knowledge, considering mobile phone use in small, like-minded networks has been linked to less involvement in other areas of political life. That is, the negative implications of monadic clusters found in previous research on political involvement does not translate to decreased levels of policy knowledge, at least not during the 2008 US presidential election.

With these collective insights as a foundation, we now reflect on the implications of mobile communication and social networks for the social policy environment more broadly. The policy environment cuts through various levels of social order. At the top, our elected officials debate and make policy through legislation. Policy is also debated and shaped at the grassroots level. The notion of the *public* sphere (Habermas, 1989) highlights the importance of discourse that takes place in the *private* realm of social relations. Whether it take place in public areas such as salons and coffee houses or through mediated channels such as the mobile phone, free and open discussion of societal problems among autonomous citizens plays an important role in the shaping of social policy. Through discourse in the private realm of social relations, opinions about social policy are formed (and informed), while voices are both expressed and heard. When these voices culminate into collective action, elected officials take notice and, ideally, take them into consideration when making social policy.

However, there may be an important distinction between mobile-mediated discourse and that which takes place with co-present others in public areas. That is, the rise of mobile communication may be fueling a trend toward the privatization of policy discourse. Mobile-mediated discourse is more private and selective than interactions that take place in a salon, coffee shop, or Web forum, which provide opportunities for others to enter the discussion. Mobile communication has come to be embraced as a means for discussing politics and public affairs, especially during the 2008 US presidential election (Aleo-Carreira, 2008; AT&T, 2008; Campbell & Kwak, in press). While this added layer of communication has the potential to support political knowledge and engagement, it may also be shifting the balance between discourse in open settings and private exchanges among individuals who know each other well. Gergen argues this cultural transformation is underway, to the extent that the very existence of autonomous individuals as the foundation of civil society is threatened by the formation of monadic clusters. Not only can these clusters hinder participation and open debate, they may also have a balkanizing effect that hampers collective action and the transcendence of voices to higher levels of the social policy environment. Although there were no significant findings for the three-way interaction between size, like-mindedness, and mobile communication, the significant (or marginally significant) findings for the two-way interactions suggest that intense mobile communication in networks that are either small or like-minded may not cultivate a constructive discussion context for the policy environment.

At the surface, these concerns may appear to simply be an extension of the ongoing debate over the individualizing effects of communication technology. For example, Putnam (2000) warns about individualistic retreat from public life through television viewing, while others have expressed similar concerns about the Internet (Nie & Erbring, 2002; Nie & Hillygus, 2002). However, concerns about mobile communication are not so much associated with individualism as they are with social privatism (Fischer, 2005; Gergen, 2008; Wilken, 2010). Social privatism is distinct in that individuals are still socially connected. They may be turning inward, but not entirely into the self as they maintain strong connections to close friends and family members. One might consider social privatism to be less worrisome than individualism because people are not isolated, and social capital is generated through connections with close personal ties (Ling, 2008; Wilken, 2010). However, immersion in one's private sphere of close friends and family through anytime, anywhere connectedness may still have deleterious consequences for political life when those network ties are very few in number and tend to reinforce each other's views. As noted, the flip side of this coin is that having access to these ties through the mobile phone may enhance discourse, knowledge, and engagement when networks contain constructive properties. Rather than simply helping or hindering civil society, it is more accurate to posit that mobile communication tends to amplify trends associated with social network ties, be they positive or negative. However, this tendency varies depending on which aspect of political life is being examined. While mobile communication in larger networks has been linked to a marked increase in engagement in previous research (Campbell & Kwak, in press), this study shows only a very slight increase in policy knowledge under these circumstances.

While this study uncovers trends for a prominent, yet still fairly new, channel for social networking and political communication, it is important to acknowledge that the survey was cross-sectional, and therefore we are unable to make definitive claims about causality. Theoretically, it is highly conceivable that policy knowledge is affected by mobile discussion among close ties. However, it is also possible that knowledge has an impact on how often one uses the technology, particularly for discussing politics and public affairs. Teasing out the direction of causality is an important next step in this line of work. In developing a theoretical model, it will also be useful to examine whether mobile communication influences social network characteristics or whether that causal arrow points in the other direction, which is very possible considering various uses the technology are shaped by one's close network ties (Campbell & Russo, 2003).

Future research in this area should also strive for a nuanced account of different user segments. As Gergen (2008) qualifies, it is quite possible monadic clusters represent a particular niche in civil society. Because existing research demonstrates notably different mobile communication trends in different age groups (e.g., Castells, Fernández-Ardèvol,

Qiu, & Sey, 2007; Goggin & Crawford, in press; Ling, 2004), as well as age-related differences in political attitudes and behavior (Delli Carpini, 2000; Quintelier, 2007), segmentation studies should start by using age and life stage as an approach for identifying user groups.

As a next step we also suggest looking at differential usage patterns. Existing research on both mobile and computer-mediated communication shows that how one uses the technology plays an important role in benefits gained (Campbell & Kwak, 2010; Shah, Kwak, & Holbert, 2001). Notably, mobile communication for information exchange has been linked to increased political participation, while its use for social connections of a personal nature was not, showing that *how* one uses the technology can be just as important as *how much* when considering its implications for political life (Campbell, & Kwak, 2010). Findings from this study highlight the importance of considering *with whom* the technology is used as well.

The evidence from this study, as well as the others discussed in this chapter, indicate both promise and peril for the policy environment in the recently "mobilized" society. Increased use of the technology for political discussion with strong ties has now been linked in both positive and negative ways with policy knowledge, political participation, and willingness to discuss alternative views. The direction of these associations depends on the nature of one's network of close ties, suggesting a synergistic relationship between mobile technology and social networks. Theory building in this area should account for this trend in the findings, while looking further into the causal flow of associations between mobile communication, network characteristics, and civic life.

References

Aleo-Carreira, C. (2008, November 5). Did social media decide the U.S. presidential election in Obama's favor? [Online exclusive]. *The Industry Standard.* Retrieved from http://www.thestandard.com/news/2008/11/05/did-social-media-decide-u-s-presidential-election-obamas-favor.

Arendt, H. (1968). Truth and politics. In H. Arendt (Ed.), *Between past and future: Eight exercises in political thought* (pp. 227-264). New York: Viking Press.

AT&T. (2008, November 6). *AT&T reports largest text messaging spike in company history on election night* [Press release]. Retrieved from http://www.att.com/gen/press-room?pid=4800&cdvn=news&newsarticleid=26287.

Benhabib, S. (1992). *Situating the self.* New York: Routledge.

Campbell, S. W., & Kwak, N. (2010). Mobile communication and civic life: Linking patterns of use to civic and political engagement. *Journal of Communication, 60(3), 536-555.*

Campbell, S. W., & Kwak, N. (in press). Political involvement in "mobilized" society: The interactive relationships among mobile communication, network characteristics, and political participation. *Journal of Communication.*

Campbell, S. W., & Kwak, N., Bae, S-Y., Brown, K., Lee, H., & Yu, R. (2010, August). Mobile communication and the personalization of public life: Implications for open dialogue. Paper presented at the annual meeting of the Association for Education in Journalism and Mass Communication, Denver.

Campbell, S. W., & Russo, T. C. (2003). The social construction of mobile telephony: An application of the social influence model to perceptions and uses of mobile phones within personal communication networks. *Communication Monographs, 70*(4), 317-334.

Castells, M., Fernández-Ardèvol, M., Qiu, J, & Sey, A. (2007). *Mobile Communication and Society: A Global Perspective*. Cambridge, MA: MIT Press.

Daft, R. L. & Lengel, R. H. (1986). Organizational information requirements, media richness and structural design. *Management Science 32*(5), 554-571.

Delli Carpini, M. X. (2000). Gen.com: Youth, civic engagement, and the new information environment. *Political Communication, 17*(4), 341-349.

Eveland, W. P., & Hively, M. H. (2010). Political discussion frequency, network size, and "heterogeneity" of discussion as predictors of political knowledge and participation. *Journal of Communication, 59*(2), 205-224.

Fischer, C. S. (2005). *Bowling Alone*: What's the Score? *Social Networks, 27*(2), 155-167.

Fulk, J., Schmitz, J., & Ryu, D. (1995). Cognitive elements in the social construction of technology. *Management Communication Quarterly, 8*(3), 259-288.

Fulk, J., Schmitz, J., & Steinfeld, C. W. (1990). A social influence model of technology use. In J. Fulk & C. Steinfeld (Eds.), *Organizations and communication technology* (pp. 117-139). Newbury Park, CA: Sage.

Gergen, K. J. (2008). Mobile communication and the transformation of the democratic process. In J. Katz (Ed.), *Handbook of mobile communication studies* (pp. 297-310). Cambridge, MA: MIT Press.

Goggin, G. & Crawford, K. (in press). Generational disconnections: Youth culture and mobile communication. In R. Ling & S. Campbell (Eds.), *Mobile communication: Bringing us together and tearing us apart*. New Brunswick, NJ: Transaction Publishers.

Habermas, J. (1989). *The structural transformation of the public sphere: An inquiry into a category of bourgeois society*. Cambridge, MA: MIT Press.

Habuchi, I. (2005). Accelerating reflexivity. In M. Ito, D. Okabe, & M. Matsuda (Eds.), *Personal, portable, pedestrian: Mobile phones in Japanese life* (pp. 165-182). Cambridge, MA: MIT Press.

Katz, J. E. (Ed.) (2008). *Handbook of mobile communication studies*. Cambridge, MA: MIT Press.

Katz, J. E., & Aakhus, M. A. (Eds.) (2002). *Perpetual contact: Mobile communication, private talk, public performance*. Cambridge, UK: Cambridge University Press.

Kwak, N., Williams, A., Wang, X., & Lee, H. (2005). Talking politics and engaging politics: An examination of the interactive relationships between structural features of political talk and discussion engagement. *Communication Research, 32*(1), 87-111.

Ling, R. (2004). *The mobile connection: The cell phone's impact on society*. San Francisco: Morgan Kaufman Publishers.

Ling, R. (2008). *New tech, new ties: How mobile communication is reshaping social cohesion*. Cambridge, MA: MIT Press.

Manin, B. (1987). On legitimacy and political deliberation. *Political Theory, 15*(3), 338-368.

Matsuda, M. (2005). Mobile communication and selective sociality. In M. Ito, D. Okabe, & M. Matsuda (Eds.), *Personal, portable, pedestrian: Mobile phones in Japanese life* (pp. 123-142). Cambridge, MA: MIT Press.

McLeod, J. M., Scheufele, D. A., & Moy, P. (1999). Community, communication, and participation: The role of mass media and interpersonal discussion in local political participation. *Political Communication, 16*(3), 315-336.

McPherson, M., Smith-Lovin, L., & Brashears, M. (2006). Social isolation in America: Changes in core discussion networks over two decades. *American Sociological Review, 71*(3), 353-375).

Mutz, D. C. (2002a). Cross-cutting social networks: Testing democratic theory in practice. *American Political Science Review, 96*(1), 111-126.

Mutz, D. C. (2002b). The consequences of cross-cutting networks for political participation. *American Journal of Political Science, 46*(4), 838-855.

Nie, N. H., & Erbring, L. (2002). Internet and society: A preliminary report. *IT & Society 1*(1): 275-283. Retrieved from http://www.stanford.edu/group/siqss/itandsociety/v01i01/v01i01a18.pdf.

Nie, N. H., & Hillygus, S. D. (2002). The impact of Internet use on sociability: Time-diary findings. *IT & Society, 1*(1), 1-20. Retrieved from http://www.stanford.edu/group/siqss/itandsociety/v01i01/v01i01a01.pdf.

Putnam, R. D. (2000). *Bowling Alone: The Collapse and Revival of American Community.* New York: Simon & Schuster.

Scheufele, D. A., Hardy, B.W., Brossard, D., Waismel-Monor, I., & Nisbet, E. (2006). Democracy based on difference: Examining the links between structural heterogeneity, heterogeneity of discussion networks, and democratic citizenship. *Journal of Communication, 56*(4), 728-753.

Scheufele, D. A., Shanahan, J., & Kim, S. (2002). Who cares about local politics? Media influences on local political involvement, issue awareness, and attitude strength. *Journalism and Mass Communication Quarterly, 79*(2), 427-444.

Shah, D. V., Kwak, N., & Holbert, R. L. (2001). "Connecting" and "disconnecting" with civic life: Patterns of Internet use and the production of social capital. *Political Communication, 18*(2), 141-162.

Quintelier, E. (2007). Differences in political participation between young and old people. *Contemporary Politics, 13*, 165-180.

Wilken, R. (2010). Bonds and bridges: Mobile phone use and social capital debates. In R. Ling & S. Campbell (Eds.), *Mobile communication: Bringing us together or tearing us apart?*. New Brunswick, NJ: Transaction Publishers.

8

Mobile-mediated Publics in
South Africa's 2009 Elections

Marion Walton and Jonathan Donner

This chapter investigates mobile mediated political participation during the 2009 national elections in South Africa, which reflected two profound changes in the global communications environment: the widespread adoption of mobile telephony in the global south (Donner, 2008), and the rise of social media (blogs, YouTube, Flickr, Facebook, Twitter, etc) known collectively as "Web 2.0" (O'Reilly, 2005).

Each change has implications for political participation. Consider two archetypes: First, during the text message revolutions in the Philippines, a relatively poor population took to the streets, brandishing $30 phones, and used tiny 160 character text messages in an effort to take down a government (Paragas, 2003; Rafael, 2003; Rheingold, 2002). Second, during the 2008 presidential elections in the United States, a relatively wealthy population took to their desktops and laptops, uploading mashed-up videos, organizing meet-ups, fundraising, and "friending" each other on bandwidth-heavy social networking sites (Smith & Rainie, 2008).

These archetypes exaggerate the technological divide between the global north and south. Mobile social software platforms are emerging that suit the low-bandwidth environments and cost-constraints of millions of mobile-centric Internet users in the developing world (Kolko, Rose, & Johnson, 2007). Diverse traditions of political participation and levels of mediatization differentiate north from south, but newly affordable technologies allow grassroots mobile networks to interconnect with the globally networked publics of the Internet (Benkler, 2006).

The South Africa case demonstrates how mobile social architectures can both facilitate and curtail aspects of political participation. Some systems, notably viral and broadcast SMS messages, are ephemeral and virtually invisible to non-participants. Other systems, like mobile web-sites, are flexible, public, discoverable, and archivable, but are distant from everyday spaces of deliberation in most developing countries, where they also present severe constraints on authorship for mobile-centric users. Still others are hybrid systems—mobile social applications that attract large numbers of regular users, often through instant messaging or other transient genres of talk.

The 2009 elections in South Africa show that these hybrid social net-works, while opening access with one hand, bar certain kinds of partici-pation with the other. They are not all equally accessible to marginalized communities or subaltern counterpublics (Fraser, 1992) struggling for social and discursive power. Members of such counterpublics are adopt-ing mobile social networks (as they formerly adopted other media (Haupt, 2008)) both for play and subversion. Nonetheless analysis of the 2009 elec-tions suggests that these mobile social networks did not facilitate political (at least electoral) communication, either between counterpublics and a local mediatized public sphere or globally with other networked publics. They did thus not allow for broader contestation, or for deliberation.

South Africa

South Africa, which still suffers stark economic inequalities, is home to a vibrant startup/technology community, and is capable of rapid tech-nological innovation. These factors, combined with high levels of mobile use, make it a particularly fruitful environment for the development and adoption of mobile social software different from both bandwidth-hungry "web 2.0" applications and the universal, basic SMS.

The Telecom Environment

Most of South Africa's fixed-line Internet users belong to a prosper-ous elite. In 2008 there were an estimated 4.1 million Internet users, just under 10 percent of the population (ITU, 2009). Relatively scarce undersea connections link Africa and the rest of the world. Consequently bandwidth (whether broadband or dial-up) is expensive for households, and is generally paid for by the megabyte, rather than in the monthly all-you-can-surf plans on offer in the US.

By contrast, the mobile phone has become nearly ubiquitous. Be-tween 2003 and 2008, the number of mobile subscriptions per 100 people climbed from 35.9 percent to 90.16 percent (ITU, 2009), while in the same period, landline subscriptions actually declined, from 10.3 percent to 9.2 percent. As mobile penetration has increased, so has the sophistication of handsets and the pervasiveness of mobile Internet. In 2008, South Africa was sixth on the list of countries served by AdMob's mobile advertising (AdMob, 2008).

Rather than, or in addition to using computers, many people access digital media and the Internet via their mobile phones (Donner & Gitau, 2009; Kreutzer, 2009). Kreutzer's survey of students in low-income high schools in Cape Town (2009) suggests that many young, urban South Africans first access the Internet via their phones. Many are unaware that they are using the Internet when using mobile instant messaging services or accessing operator content via GPRS.

A downloadable mobile instant messaging and chat application called MXit is market leader in South Africa. According to email from a MXit representative, it boasts 15 million registered users, 13 million of whom are South African (Laura Hallam, personal communication, 14 August 2009). In other words, many more South Africans use MXit than use the traditional PC-based Internet. Often, "first time" Internet use is mobile, and via MXit.

MXit earns revenue from advertising, and from selling premium content through micropayments in a virtual currency. Advertisers can purchase impressions on the opening "splash screen," and can pay for chat rooms where they can engage users in their campaign.

Teens first began using MXit *en masse* as a cost-saving substitute for text messages, since messages sent between individual contacts are free, other than the minimal cost of the data. Using these networks, teens find private spaces for self-reflection, friendships, and relationships away from parental supervision (Bosch, 2008). Demographics are shifting, though, and despite occasional "moral panics" among concerned parents and authorities (Chigona & Chigona, 2009), MXit now attracts a wider range of users, enjoying top-of-mind awareness in the South African social mediascape (Lombard, 2009).

The Election

April 2009 saw South Africa's fourth national elections since the country's first democratic elections in 1994. Though it was a heated contest, the outcome of the 2009 elections was never really in doubt. The ruling African National Congress (ANC) maintained its lead, winning 65.9 percent of the national vote (it received 62.6 percent in 1994, 66.4 percent in 1999, and 69.7 percent in 2004), and a majority in eight of nine provinces. This was nonetheless the first time since 1994 that the percentage of voters supporting the party had dropped. Some voters shifted their support to a new party, the Congress of the People (Cope), which received 7.4 percent. The official opposition, the Democratic Alliance (DA) grew from a low base, winning 16.7 percent of the vote, though remaining a party of the middle class and ethnic minorities.

Observers speculate that the ANC's electoral campaign cost R400-R500 million (approximately $53-$67 million) (Butler, 2009, p. 74). The party leveraged a national network of branches and mobilized hundreds of thousands of members and volunteers (Butler, 2009, p. 73). The campaign combined sophisticated marketing strategies, door-to-door campaigning, massive rallies, and "community hall" meetings. It retained traditional support bases among the rural poor, grew its support in the province of KwaZulu-Natal, and held onto many voters from impoverished urban townships, all despite deepening inequality and pervasive poverty, sluggish service delivery, evidence of government corruption, the global economic crisis, and the dramas of the ANC's internal power struggles.

Since 2005, a faction of the ANC associated with former President Thabo Mbeki had been in conflict with another faction coalescing around his former deputy, the embattled Jacob Zuma. Zuma, a leader of humble origins, became standard-bearer of the left, in particular of the Congress of South African Trade Unions (COSATU) and the South African Communist Party (SACP) and enjoyed considerable popular support despite facing charges of corruption and bribery. He was swept to power at the ANC's national conference at Polokwane in December 2007, by a broad anti-Mbeki coalition, which included impoverished delegates as well as disgruntled provincial leaders, businessmen, and the ANC Youth League (ANCYL). In response to Zuma's election, and to Mbeki's ousting from office in September 2008, Cope, a new breakaway opposition party, was formed. Cope appeared to have the potential to win the support of a large number of voters, particularly from the growing black middle class. The ANC's campaign, galvanized by Cope's appearance, focused on registering new young voters, mobilizing traditional supporters, and ensuring that voters turned out in their numbers to vote in the largely peaceful elections.

The poor hold the power in South African elections (Lefko-Everett, Misra-Dexter, & Sylvester, 2009), since about half of the population still live in poverty (Southall, 2009) and most poor black voters kept faith with the ANC. The ANC campaign signified modernity as well as continuity and successfully targeted young people. Eleven percent of the electorate were new voters, 1.3 million of them young people below 30 who were voting for the first time (February, 2009, p. 48; Lefko-Everett, et al., 2009). Subject to increasingly cosmopolitan influences through globalizing media and technologies, this "cell phone generation" were believed to be less susceptible to the characteristic South African patterns of ethnic and racial "identity voting" (Southall, 2009, pp. 9-10), and their participation had declined in previous elections (February, 2009; Feburary, 2009). While the ANC's history as liberation movement might not have weighed as heavily with these young voters as it did with their parents, the vibrant ANC campaign nonetheless succeeded in winning many of their votes, as did the sense of a party once again committed to addressing the problems of the poor.

Mobile Social Media and the 2009 Election

We introduce four elements of mobile mediated political communication that played a role in the election. Two—SMS and mobile web sites—represent the low and high-content extremes currently offered by the mobile channel. The other two—MXit and Mig33—represent low-bandwidth, mobile-specific social software applications. Although less familiar to readers outside South Africa, these two hybrid applications illustrate different approaches to mobile mediated political participation.

SMS Wars and the Tussle for Power in the ANC

A tussle for leadership of the African National Congress (ANC) dominated South African politics in the pre-electoral period 2006 and 2007, and culminated in the party's national conference in Polokwane (16-20 December, 2007), where Mbeki and Zuma were both nominated for the position of ANC party president. The conference environment was highly charged, and, at one stage, an altercation over a system of computerized voting took center stage, fuelled by a "war" of SMS messages.

Conference proceedings were brought to a halt when the ANCYL, who supported Zuma, brought a motion to disallow computerized counting of votes, demanding a manual count (Lund, 2007). ANCYL delegates apparently feared that Mbeki and his supporters, who had greater access to the resources of government, could use the automated vote-counting system to skew the results in their favor. They spread SMS messages claiming that the vote-counting program had been manipulated to Mbeki's advantage (Monare, 2007).

The opposing camp was equally adept at using SMS messages to influence events during the conference. Another rumor, likely originating from the Mbeki camp, claimed that a company associated with the brother of Zuma's disgraced financial advisor had been contracted to handle and secure ballot papers. "More gost voters registered, hence insistence on manual counting. It's over. The ANC has bin hijacked," (quoted in (Monare, 2007)). MW: Yes…authentic SA txtspk… didn't want to use 'sic' as that is v patronizing.

The software was not used to count the presidential votes, and Zuma was elected party president. Moreover, a "Zuma list" of candidates for the top six offices circulated via SMS and led to the defeat of a whole swathe of Mbeki loyalists vying for top positions (Butler, 2009, p. 69). Similar text wars characterized a series of intra-ANC dramas until the party leadership ousted Mbeki from the presidency of South Africa in September 2008, just seven months before the elections.

".mobi is the Way Forward"

> ANC Update: Now is the time for us to focus on improving people's lives, Jacob Zuma says, after NPA drops charges. For the full response go to www.myanc.mobi.

Shortly before the National Election in 2009, the ANC sent this message to supporters who had signed up to receive updates via their mobiles. It marked a landmark in the campaign, announcing that the National Prosecuting Authority had dropped criminal charges against Zuma, who was by then the ANC's presidential candidate. It also encouraged supporters to visit the www.myanc.mobi website, a newer mobile version of the ANC's "traditional" social media portal www.myanc.org.za. The .mobi site encouraged visitors "Have your say. We are listening". One supporter posted the following message:

> ANC keep up all the gud work, keep delivering. Aluta continua! .mobi is the way forward.

Mobile usage was lower than campaigners had hoped. According to Steyn Speed (2009), then the ANC's acting national spokesperson, the mobile site attracted less traffic than the main campaign site, myanc.org.za (which had attracted 50,000 cumulative visits through May 2009). The ANC nevertheless perceived mobile to be important:

> The political party or social institution that manages to use cell phones to communicate around the issues that affect people is going to reap the benefits… we still need to find out how (Speed, 2009).

The ANC social media campaign on myanc.org.za and on Facebook, YouTube, 24.com, blogs, and Twitter aimed primarily "to engage with people who aren't necessarily going to vote for [the ANC]" (Speed, 2009). The Facebook friending, blog-posting, tweeting population are influential in shaping elite online opinion, but South Africa's roughly

4 million "traditional" Internet users are not the ANC's primary constituency. To ensure a wider digital reach a mobile site was essential.

Contributors to the .mobi site declared their loyalty and love for the party, but also demanded action on issues such as unemployment and education, which concern voters, but were neglected in party manifestos and event-driven, personality-focused election coverage (Duncan, 2009). Contributors criticized abuses, bewailed slow service delivery ("Areas under our control for 15 years without service delivery shame the organization") and volunteered their services to the organization. The unexpected threat posed by Cope formed a strong rallying point, as did attacks on the DA.

In contrast, the DA focused on building a complex, "Obama-esque" campaign website. Their mobile site (mobi.da.org.za) was marginal to the campaign. It neither incorporated any social features, nor made use of text-messaging to reach the majority of South African voters (Buckland, 2009).

Thus mobile-centric users were excluded from participating in the DA campaign, unless they could afford the high bandwidth costs of the mobile version of Facebook. Neither the DA nor other political parties were able to mobilize significant levels of Obama-style online participation via Facebook or their own social media sites, since their strategies focused on under-utilized conventional websites, rather than localizing social media strategies through mobile communication (Duncan, 2009).

Of the official campaign sites, the ANC .mobi site went furthest in encouraging mobile users to participate, but while the site assured users "we are listening," officials did not respond to comments on the site, even when supporters posted direct questions. In contrast, as we will show, an unofficial campaign by the ANC's Youth League used a localized version of social networking to allow mobile interaction with national leaders and to shift patterns of mobile use among its national network of supporters, albeit briefly.

MXit Sits It Out

Given MXit's prominence in the South African mobile communications landscape, one might expect to see it playing a role in political parties' election campaigns, much as Facebook and YouTube had featured so prominently in the Obama and McCain campaigns in the US.

But when the ANC approached MXit to host their campaign, they were refused (Speed, 2009). A MXit representative explained in an email to us that MXit was "purely an application for communication" and did not want to be seen to endorse any particular party's policies (Laura Hallam, 2009, personal communication, 31 July 2009). MXit also sent us a copy (via email) of their content policy, which forbids "content containing religious, political and/or common social issues, such as abortion and suicide" ("Content policy [external policy]—Annexure," 2009, personal communication, 31 August 2009). Speed reconfirmed MXit's decision in an SMS to us: "MXit said they didn't want political parties on their platform" (Steyn Speed, personal communication, 18 August 2009).

The MXit architecture allows MXit to handpick the messages promoted on their platform, and this, together with the size of their audience, gives them considerable editorial clout. At a meeting with us, MXit's international marketing manager explained that their editorial decisions prioritize youth-focused and interactive campaigns (Juan du Plessis, personal communication, 7 September 2009). So, for example, MXit turned down Jacob Zuma's first Presidential State of the Nation address, but allowed other political campaigns, such as Youth Day, and sent messages from users to US President Barack Obama during his 2009 trip to Africa. Given recent media panics about the safety of MXit's young users (Chigona & Chigona, 2009), perhaps MXit Lifestyle didn't want to wander into possibly controversial territories at the time of the elections. Whatever the rationale, the absence of politically themed chatrooms on MXit on the eve of the election (or at any other time) is significant to those who look to mediated mobile social networks as new arenas for emergent publics.

Unlike recent censorship activities in Iran and China, no government officials blocked political content on MXit, nor did MXit censor individual chats between its users. However, to echo Cohen (1963) and McCombs and Shaw (1972), while MXit's actions may not have told its users what to think, they helped set the agenda by suggesting what they should think *about*. At the time of the elections, MXit's chatrooms were organized around the age and location of the users (e.g., "teens" or "Cape Town"), while other rooms encouraged users to explore interests defined primarily by consumption such as "technology," "fashion" and "cars." MXit does not allow user-generated content other than user-uploaded music and classified advertisements.

Most DA supporters and the well-heeled echelons of the ANC could turn to Facebook to express themselves politically. But many MXit users would be unlikely to know of that option, even if they could afford the bandwidth. By excluding politically themed chats and user-generated

content, MXit's content policy curtailed opportunities for public participation on the most popular "new media" platform in the nation.

The Mig33 Campaign and Mobile Mass Meetings

With the biggest player, MXit, on the sidelines, another mobile social
network grasped the opportunity to host political debates and promote
itself among mobile-savvy South African voters. Based in California,
Mig33 is a mobile application which reports 20 million registered users
(Marshall, 2009) of which a few million are South African (Engelbrecht,
2009). Unlike the mobile version of Facebook, Mig33 supports real-time
interaction with instant messaging, and caters to phone users with text
messages, VoIP, and a low-bandwidth design. Unlike MXit, Mig33 allows
individual users to create public chatrooms where they can meet people
with similar interests to discuss any topic of their choice, and offers users
the option to craft public, searchable profiles. Another feature, unique
to Mig33, is the "stadium," supporting mass chats between leaders and
up to 5000 supporters.

Mig33 invited South African political parties to use Mig33 to set
up groups that would allow mobile campaigning (Engelbrecht, 2009).
Mig33 allows groups to communicate asynchronously with their members through announcements, and to display lists of registered members'
profiles.

The ANC, DA, Cope, and two smaller parties set up Mig33 groups.
Of these, the most active users were the ANC's Youth League and Cope,
although Cope's use was more sporadic, and their Mig33 group was
labelled an "unofficial site." The ANCYL and their alliance partners sent
out notices in emails and SMS messages (e.g., (Ngobese, 2009) asking
members to install Mig33. After having joined, members received notices
about upcoming events, both virtual gatherings such as appearances by
ANCYL leaders in the Mig33 chat "stadium," and actual meetings such as
braais (known elsewhere as BBQs, grills, or cookouts) and rallies. While
not part of the official campaign, the group was carefully integrated with the
ANCYL's overall campaign. The Mig33 chat rooms proved considerably
more popular than the ANC's .mobi site, possibly because young ANCYL
members were accustomed to browsing the Internet on their phones, and
were experienced in the practices of instant messaging and MXit chat.

Over a few months, the Mig33 ANC group had over ten thousand
registered members, and the ANCYL had used the group to post 37 announcements. These notices included commentary on political events,
announcements of chats with ANCYL leaders in Mig33's chat stadium,

and messages designed to mobilize large crowds (e.g., "ANCYL Katle-hong Ride Braai on 05 April show your support"). Mig33 was not used by the majority of ANC members, but it appears to have succeeded in engaging a core group of young ANCYL activists and supporters during the election period. To put the numbers in context, only 700 members had joined the Facebook ANCYL group. That said, the actual number of ANC supporters on Mig33 is difficult to ascertain. On the one hand, many non-South Africans joined the group and there were also some "spam" accounts. On the other hand, some supporters joined chats and not the group. In the weeks before the elections, there were usually one or two full ANC chatrooms.

Few members of the Mig33 group created profiles, perhaps indicating that Mig33 was a temporary supplement to their regular social network-ing practices, (possibly on MXit, or on Facebook). This possibility is borne out by the fact that after the election results were announced, the ANCYL use of the Mig33 group dwindled almost immediately. While a few notices were posted by the ANCYL leader, Julius Malema, the chatrooms emptied, and the site bore no traces of the lively debates, bon-homie, and dramatic conflicts which entertained, enraged and enthused participants during the election.

Discussion

Popular narratives of the role of mobile technology in the politics of developing countries tend to emphasize moments of popular indignation transformed into mass mobilization (e.g., in (Rheingold, 2002). Mobile devices are sometimes given too much credit for the outcomes of mass mobilizations (Miard, 2009; Rafael, 2003). Other factors also shape these mass mobilizations, such as when, for example, initial SMS reports of SARS were squashed by government-controlled mass media in China (Castells, Fernández-Ardèvol, Qiu, & Sey, 2007). Mobile use in the 2009 South African elections certainly does not fit the indignation/revolution model that has dominated discussion to date. The South African case sug-gests that we should notice "everyday" as well as exceptional events, and that privately owned new media platforms as well as governments and populace are political players. This will allow a more nuanced, if more mundane, view of the political affordances of mobile technology.

Mobile Social Networks as Mediated Publics

Boyd (2007) has suggested that online social networks only consti-tute a mediated public when they are characterized by four properties:

persistency, when messages are stored indefinitely, thus becoming available for later scrutiny; *searchability*, which allows convenient keyword-based access to messages by audiences other than their original recipients; *replicability*, which allows messages to be copied and used again; and *invisible audiences*, or the larger audiences who can access the records of conversations between smaller, known groups of people. In these ways, social network sites shift social interaction away from a particular moment, and make the record of that interaction asynchronous and accessible to others outside that context.

None of the major forms of mobile mediated political communication described above fully meet boyd's criteria for social networks as mediated publics. The 160 character bullets in the SMS wars at Polokwane offer none of these features. At the other extreme, the .mobi sites offered all four, but fell short of providing profiles and "friends lists."

Among the newer hybrid forms, Mig33 offered access to searchable profiles and invisible audiences. But chats were not recorded, and are neither searchable nor easily replicated, thus reducing the site's value as an archive and wellspring of reusable political content.

MXit, too, did not engender a mediated public by boyd's criteria. MXit profiles are not publically accessible, and even users who are mutual contacts cannot view one another's contact lists. As discussed above, MXit had no formal venues for political participation. While friends certainly exchanged messages about politics, and many seemingly apolitical uses of popular culture or music are associated with subaltern counterpublics (cf. Haupt, 2008), these messages and discussions on MXit were not persistent, searchable, or replicable. Nor was it possible to interconnect with a broader invisible public watching, learning, assessing or contesting.

Walled Gardens and Intermediate Spaces

Historically, public spheres have claimed to be inclusive and not to respect status differentials while they effectively restrict access and participation by marginalized groups. We see this trend across all four examples of mobile use in the 2009 South African election and their failures to produce a "networked" (Benkler, 2006) or "mediated" (boyd, 2007) public. In particular, the mobile social networks still present obstacles to full democratic access, defined as the ability to use them to organize and locate counterpublics, and to accommodate local genres of grassroots political action and expression. This suggests a few significant challenges for mobile-mediated political communication.

First, the MXit example illustrates how the interaction of "walled gardens" and closely controlled user experiences can restrict access to political participation. This is not the case with all mobile applications, as the approach of Mig33 makes clear. Rather, it reveals how an institution (in this case, the privately held MXit) can exert editorial control over its users. MXit's position, as both the first and currently only Internet experience for millions of South Africans resembles America Online's relationship with its users in the early 1990s (Patelis, 2000). Previously discernable differences between content-neutral communications providers and content-active media outlets have blurred, creating challenges for participants, analysts, and policymakers alike.

Second, the concentration of interaction on the Mig33 and MXit sites is a reminder of the distinctive genres of political communication in IM applications, as opposed to searchable, all-purpose web pages. Users were unlikely to toggle between chats and the Web, and their textual "mashups" of word of mouth and the mass media were not taken up by the mediatized public sphere. The chat on Mig33 and MXit was as fleeting as a chant or slogan heard during a street corner rally; Powerful in the moment, but not part of the public record.

Conclusion

The story of mobile Internet and South African youth mobilization around the 2009 election is primarily the story of a well-funded youth movement, the ANC's Youth League (ANCYL), whose strategies appealed to impoverished young voters and helped the incumbent ANC to win the election. The organic mobile campaign of the ANCYL routed around MXit's limitations. Largely eschewing the official ANC .mobi site, they adopted Mig33, a foreign-owned platform for their campaign.

Mobile-centric political participation was elicited enthusiastically, but this interest does not appear to have been maintained after the election. Supporters were given writing-rights, but mobile participation was not interconnected with a broader mediated public sphere, and splintered off to a separate .mobi site and transient chatrooms instead. The task of generating more sustained and democratic mobile participation will be vital for the emergence of a fully-fledged "networked" (Benkler, 2006) and "mediated" (boyd 2007) South African public.

MXit is a key actor in the local mobile Internet space and they have made low-cost interpersonal communication accessible to millions. Designed around low bandwidth users with a very low price tolerance and minimalist handsets, MXit makes a contribution of the kind that other social networks are unlikely to prioritize. Nonetheless, their editorial policies raise several important concerns. Here, an asymmetrically structured content distribution network has developed a parallel one-to-many structure for (paid) 'content', and a many-to-many structure for (free) IM. This architecture does not provide opportunities for user-generated content and other forms of public discourse and limits the range of political expression available to precisely those citizens and organizations who do not have many other ways of reaching national or international audiences. While highly motivated activists may be able to route around MXit's current limitations, as the ANCYL and Cope did when they adopted Mig33, network effects will mean that ordinary citizens are less likely to join them. If Facebook had banned Obama and McCain, what would have been the reception in the US?

For the broader and ongoing discussion about the evolution of an online mediated public sphere (Benkler, 2006; boyd, 2007), the example of the 2009 South African election illustrates both the promise and limitations of current mobile-centric platforms. It demonstrates that mobile media offer more than the raw power of the "smart mob" to draw a physical crowd. The extension of the ANC campaign to myanc.mobi and Mig33 illustrate the promise of extending online political participation and deliberation to a broader public. At the same time, it exemplifies how such counterpublics may be mobilized merely to amplify a brief political campaign rather than to ensure ongoing and higher levels of accountability. In particular, it suggests the distance between such mobile spaces and social network sites as accessible, archivable, and searchable mediated publics.

South Africa is an emerging twenty-first century democracy. Each election cycle is a milestone, and requires the country to find its footing anew, with a young and often frustrated population. Whether policy interventions are needed to open up bounded mobile social media systems, or otherwise encourage mobile social software systems to develop spaces for public discourse is a matter for further discussion in South Africa, and defies an easy description or solution. The stakes are large, but so are the difficulties associated with attempts to shape the nature of public interactions on these new platforms. Perhaps the rapid pace of technological change and the pressure of competing systems such as The Grid, Noknok, Facebook and Mig33 will ensure that soon all citizens can find spaces that can afford them writing-rights for mobile-mediated online political participation. But current limitations on participation in mobile social networks are of concern to citizens, activists, NGOs, providers, political parties, and policymakers, particularly since fixed line Internet serves so few. Even television and print media prioritize more affluent market segments, primarily the 54.1 percent of the population who are more attractive to advertisers (Duncan, 2009:298). The failures to develop a more broadly accessible mobile public sphere in South Africa's 2009 election campaign are worthy of public scrutiny, challenge and appraisal, and may require targeted steps to extend broader read-write access to the networked publics and counterpublics promised by Web 2.0.

References

AdMob. (2008). Admob Mobile Metrics Report 10 October. Retrieved from http://www.admob.com/marketing/pdf/mobile_metrics_aug_08.pdf.

Benkler, Yochai. (2006). *The Wealth of Networks: How Social Production Transforms Markets and Freedom*. New Haven: Yale University Press.

Bosch, Tanja. (2008). Wots ur ASLR? Adolescent girls' use of MXit in Cape Town. *Commonwealth Journal of Youth Studies, 6*(2).

boyd, danah. (2007). Social network sites: public, private, or what? *Knowledge Tree, 13*.

Buckland, Matthew. (2009). SA political party launches impressive, Obama-esque campaign site 20 February. Retrieved from http://www.matthewbuckland.com/?p=609.

Butler, Anthony. (2009). The ANC's National Campaign of 2009: Siyanqoba! In Roger Southall & John Daniel (Eds.), *Zunami: The South African Elections of 2009* (pp. 65-84). Sunnyside, South Africa: Jacana & Konrad Adenauer Stiftung.

Castells, Manuel, Fernández-Ardèvol, Mireia, Qiu, Jack Linchuan, & Sey, Araba. (2007). *Mobile Communication and Society: A Global Perspective (Information Revolution and Global Politics)*. Cambridge, MA: MIT Press.

Chigona, Agnes, & Chigona, Wallace. (2009). MXit up in the media: media discourse analysis on a mobile instant messaging system. *The Southern African Journal of Information and Communication 9*(2), 42-57.

Cohen, Bernard Cecil. (1963). *The Press and Foreign Policy*. Princeton: Princeton University Press.

Donner, Jonathan. (2008). Research approaches to mobile use in the developing world: A review of the literature. *The Information Society, 24*(3), 140-159. doi: doi:10.1080/01972240802019970.

Donner, Jonathan, & Gitau, Shikoh. (2009). *New paths: exploring mobile-centric internet use in South Africa*. Paper presented at the Pre-Conference on Mobile Communication at the Annual Meeting of the International Communication Association, Chicago. http://lirneasia.net/wp-content/uploads/2009/05/final-paper_donner_et_al.pdf.

Duncan, Jane. (2009). Desperately seeking depth: The media and the 2009 election. In Roger Southall & John Daniel (Eds.), *Zunami: The South African Elections of 2009* (pp. 215-231). Sunnyside, South Africa: Jacana & Konrad Adenauer Stiftung.

Engelbrecht, Lezette. (2009). Mig33 platform drives political dialogue Available. *ITWeb* 2 April. Retrieved 3 August, 2009, from http://www.itweb.co.za/sections/telecoms/2009/0904021042.asp?A=COV&S=Cover.

February, Judith. (2009). The electoral system and electoral administration. In Roger Southall & John Daniel (Eds.), *Zunami: The South African Elections of 2009* (pp. 47-64). Sunnyside, South Africa: Jacana & Konrad Adenauer Stiftung.

February, J. (2009). The electoral system and electoral administration. In Roger Southall & John Daniel (Eds.), *Zunami: The South African Elections of 2009* (pp. 47-64). Sunnyside, South Africa: Jacana & Konrad Adenauer Stiftung.

Fraser, Nancy. (1992). Rethinking the Public Sphere: A Contribution to the Critique of Actually Existing Democracy. In Craig Calhoun (Ed.), *Habermas and the Public Sphere* (pp. 109-142). Cambridge, MA: MIT Press.

Haupt, Adam. (2008). *Stealing Empire: P2P, intellectual property and hip-hop subversion*. Cape Town: HSRC Press.

ITU. (2009). Online statistics, ICTeye Retrieved 11 December, 2009, from http://www.itu.int/ITU-D/icteye/Indicators/Indicators.aspx.

Kolko, Beth E., Rose, Emma J., & Johnson, Erica J. (2007). *Communication as information-seeking: the case for mobile social software for developing regions* Paper presented at the 16th international conference on the World Wide Web (WWW), Banff, Alberta, Canada.

Kreutzer, Tino. (2009). *Generation mobile: online and digital media usage on mobile phones among low-income urban youth in South Africa*. MA, University of Cape Town, Cape Town. Retrieved from http://www.tinokreutzer.org/mobile/Mobile-OnlineMedia-SurveyResults-2009.pdf.

Lefko-Everett, Kate, Misra-Dexter, Neeta, & Sylvester, Justin. (2009). Idasa 2009 Election Response 14 July. Retrieved from http://www.ngopulse.org/files/Election%20Response%202009(3).pdf.

Lombard, Charl. (2009). SA teens prefer MXit to Facebook, Google 1 June. Retrieved from http://www.itnewsafrica.com/?p=2730.

Lund, Troye. (2007). Heated dispute over voting, from http://www.fin24.co.za/articles/default/display_article.aspx?ArticleId=1518-25_2239606.

Marshall, Matt. (2009). Mig33 is most downloaded mobile app—have you heard of it? *VentureBeat* 25 June. Retrieved 4 August, 2009, from http://digital.venturebeat.com/2009/06/25/mig33-is-most-downloaded-mobile-app-have-you-heard-of-it.

McCombs, Maxwell E., & Shaw, Donald L. (1972). The agenda-setting function of mass media. *Public opinion quarterly, 36*(2), 176.

Miard, Fabien. (2009). *Call for Power? Mobile phones as facilitators of political activism'*. Paper presented at the 50th annual convention of the International Studies Association, New York. http://www.miard.ch/papers/phonesandpowerII.pdf.

Monare, Moshoeshoe. (2007). It's war by SMS at Polokwane. 18 December. Retrieved from http://www.iol.co.za/index.php?click_id=13&set_id=1&art_id=vn20071218050612824C683104.

Ngobese, Castro. (2009). Press Release: Invitation to chat with WCLSA National Secretary Buti Manamela 20 April. Retrieved from http://groups.google.com/group/yclsa-press/browse_thread/thread/a045aab36ad6fa7f?pli=1.

O'Reilly, Tim. (2005). What Is Web 2.0 Retrieved July 25, 2006, from http://www.oreillynet.com/pub/a/oreilly/tim/news/2005/09/30/what-is-web-20.html.

Paragas, Fernando. (2003). Dramatextism: Mobile Telephony and People Power in the Philippines. In Kristóf Nyíri (Ed.), *Mobile Democracy: Essays on Society, Self, and Politics* (pp. 259-283). Vienna: Passagen Verlag.

Patelis, Korrina. (2000). E-Mediation by America Online. In Richard Rogers (Ed.), *Preferred Placement: Knowledge Politics on the Web* (pp. 49-63). Maastrict: Jan van Eyck Academie.

Rafael, Vicente L. (2003). The cell phone and the crowd: Messianic politics in the contemporary Philippines. *Public Culture, 15*(3), 399-425.

Rheingold, Howard. (2002). *Smart mobs: The next social revolution*. Cambridge, MA: Perseus Books.

Smith, Aaron, & Rainie, Lee. (2008). The internet and the 2008 election. Retrieved from http://www.pewinternet.org/~/media//Files/Reports/2008/PIP_2008_election. pdf.pdf.

Southall, Roger. (2009). Zunami! The context of the 2009 election. In Roger Southall & John Daniel (Eds.), *Zunami: The South African Elections of 2009* (pp. 1-22). Sunnyside, South Africa: Jacana & Konrad Adenauer Stiftung.

Speed, Steyn (2009, 13 May). [ANC Social Media Campaign—Talk and Video at Worldwide Creative's 'Heavy Chef' Seminar, 13 May].

9

EMF Social Policy and Youth Mobile Phone Practices in Canada

Rhonda N. McEwen and Melissa E. Fritz

In this chapter we analyze Canada's social policy response to public health concerns about exposure to electromagnetic frequencies (EMF) related to mobile phone. The primary objective is to consider the factors that may or may not contribute to how Canadian policy makers interpret and enact policy with respect to EMF exposure standards. It is not the goal of this chapter to advance a point of view in the debate surrounding EMF health claims. While this is an interesting topic, existing literature covers the controversy surrounding the issue (Linder, 1995; Poole, 1990; Grasso, 1998; Storrs, 1998). Instead we utilize EMF as an example of a technological issue that has attracted controversy from a public health perspective (i.e., over 80 percent), this has not had an impact on policy agenda setting for EMF in Canada.

We consider the manner in which social policy on mobile phone related EMF is complicated by factors such as technological ubiquity, a paucity of information on mobile phone practices, and ideologies of "public protection" as articulated through exercises in social construction. Among the questions raised in our analysis we focus on the following, (a) why is there currently no strong social policy from the Government of Canada with respect to EMF, particularly as it relates to mobile phone use among young Canadians and an apparently concurrent lack of public debate?; and (b) why is there a dearth of research with respect to the information practices of youth and mobile phones in Canada?

The chapter is structured around an investigation of three explanations for the Canadian government's apparent reticence to develop a policy directive on EMF. The first possible explanation is that Health Canada, the Canadian federal ministry governing health-related matters, purportedly requires conclusive scientific evidence of harm before issuing policy guidelines or standards.

The second explanation involves folk-framing among the Canadian public that the Canadian government tends to follow the United States on policy and regulatory matters of a controversial nature. For both of these explanations, in Section 3 we provide evidence to the contrary using Health Canada's policy position on bis phenol A as an example.

A third possible reason is that Canada's low-moderate penetration of mobile phones relative to penetration in countries where concern for EMF has become prominent accounts for the lack of mobilization around the issue. Burgess (2003) states that "in the absence of a prominent mobile phone infrastructure and without the ubiquitous presence of handsets within society reaction appears less likely. The comparative lack of organized concern in Canada—a society that shares so much else with the political cultures of countries where concern has become prominent—is partly explained by the undeveloped nature of the mobile network" (p. 219). Although Burgess' statement reflected Canada's mobile phone adoption in the early part of this decade, this perception of lower mobile phone use in Canada is perpetuated today by, for example, the International Telecommunication Union statistics that do not account for the skewed ratio of population size to land mass in Canada. In Section 4 we provide empirical evidence that although mobile phone use in Canada is now significant, particularly among young people. (For the purposes of this discussion we define young people as persons between 13 and 18 years old, and children between 2 and 12 years old.)

In this chapter we offer an alternate explanation—that a lack of understanding about youth mobile phone practices couples with institutional difficulties in conceptualizing the youth population in policy frameworks, to result in slow or no policy decisions in Canada.

After a brief definition of EMF and a description of the health concerns as articulated in the scientific literature, we begin the analysis by comparing Health Canada's firm stance regarding health concerns over another recent issue, bis phenol A (BPA), to the insubstantial position adopted by the same agency with respect to EMF and mobile phones. In an effort to account for this policy asymmetry we provide some context by examining the current state of mobile phone use in Canada, and discuss youth mobile phone practices based on recent research in Toronto (McEwen, 2010). We then briefly follow the trajectory of EMF policy in other countries and compare how this issue is currently being articulated within policy institutions in Canada. By integrating

Schneider and Ingram's (2008, 2009) theory of the Social Construction of Policy with Hacker's notion of Policy Drift we consider the way Canadian youth may be viewed by policymakers as a target population in transition, which poses a challenge to policymakers given the complex and obscure structure of wireless policy and regulation in Canada. We suggest that a poor understanding of mobile phone practices couples with the complex policy structure, leading to a breakdown among policy, practice and health information in the case of EMF.

What Is EMF?

Electromagnetism is comprised of two naturally occurring fields—electrical and magnetic. Electrical fields are created by differences in voltage: the higher the voltage, the stronger will be the resultant field. Magnetic fields are created when electric current flows: the greater the current, the stronger the magnetic field. Electromagnetic fields (EMFs) occur naturally and are present everywhere in our environment but are invisible to the human eye.

However, some sources of EMF are man-made for example X-rays and domestic electricity that is used from household power sockets. While domestic electricity is considered to be a source of "super low-frequency" EMF in the 50-60 Hz range, other equipment receive and transmit information on higher frequency radio waves, for example television antennas (54-216 MHz), radio stations (AM 540-1700 kHz and FM 88.1-107.9 MHz), and mobile phones (800 MHz and 2 GHz ranges). Most mobile phones operators utilize the 800 MHz ultra-high-frequency (UHF) band, while new entrants have recently purchased the right to use spectrum in the 2 GHz range in Canada.

Concerns and Controversy about EMF

Mobile phone users are likely to have had the passing thought after a particularly warm and lengthy conversation that there may be something unhealthy about using a mobile phone close to the head. These vague concerns are reinforced by popcorn and egg cooking spoofs that went viral on popular distribution channels such as email and YouTube. While there are examples of similar fears of radiation harm with the introduction of other technologies such as television and microwave ovens (Matthes, 1992; Jensh, 1997; Burgess, 2003), the question of whether mobile telephones cause health problems remains unanswered.

In workplace environments, research has demonstrated that absorbing radio frequencies and microwave energy above certain levels can

cause harmful effects to the human body (Gandhi, Lazzi & Furse, 1996).
However, there are also studies demonstrating that the thermal effects
of EMF are insufficient to cause anatomical harm (Bennett, 1994; Higa-
shikubo, Ragouzis, Moros, Straube & Roti, 2001), and this is key since
thermal effects are the basis for EMF standards in Canada, including
those governing safety standards for EMF in mobile phones. The controversy comes from arguments from some members in the scientific
community that thermal effects are only one possible demonstration of
damage to human tissue caused by EMF, and that biological effects may
still be propagated by EMF.

In addition, research is showing a possible correlation between
EMF from mobile phones and disturbances in sleep patterns (Leitgeb,
Schrttner, Cech & Kerbl, 2008). According to Dr Chris Idzikowski, the
director of the Edinburgh Sleep Centre and one of the investigators in
a 2008 sleep study in Sweden, "there is now more than sufficient evidence, from a large number of reputable investigators who are finding
that mobile phone exposure an hour before sleep adversely affects deep
sleep." The findings were particularly concerning for children and young
people who use their mobile phones late at night and who especially
need sleep. According to the researchers, failure to get enough sleep
can lead to mood and personality changes, ADHD-like symptoms, depression, lack of concentration and poor academic performance. This
research was criticized on methodological grounds (NHS, England
2008), and along with similar types of studies the evidence of harm
remains inconclusive.

Based on increasing concerns from the scientific community and general publics globally regarding non-thermal effects of EMF, the World
Health Organization has conducted a multi-country and longitudinal
study called the Interphone Study. Results of this study are currently
pending.

Finally, researchers are increasingly concerned with negative effects
of EMF on young people. As with young people globally mobile phone
use is ubiquitous among Canadian youth. Research has demonstrated that
the biological physiology of the brains of younger children and youth
permit deeper penetration of EMF radiation than the brains of adults
(Kheifets, Repacholi, Saunders & van Deventer, 2005). Children and
youth are traditionally viewed as policy targets, with particular emphasis
on protection. Most recently, the European Parliament overwhelmingly
voted to adopt the Ries Report, with the following resolution: "[W]hereas
most European citizens, especially young people aged from 10 to 20,

use a mobile phone, an object serving a practical purpose and as a fashion accessory, and whereas there are continuing uncertainties about the possible health risks, particularly to young people whose brains are still developing."

Safety Code 6 and the City of Toronto

Regarding EMF in Canada "Safety Code 6" is the official guideline for policymakers. Health Canada has established limits of exposure to radio frequency energy based on a review of experimental evidence of detectable effects in animals and cell (i.e., biological) systems. The limits are measured and expressed in Specific Absorption Rate (SAR) units. These limits are published by Health Canada in a document called "Safety Code 6—Limits of Exposure to Radio frequency Fields at Frequencies from 10 kHz-300 GHz" (1999).

In 1999 Safety Code 6 was updated to include, for the first time, mobile phones as technological devices that could be sources of radio frequency exposure to humans. In previous version of Safety Code 6 mobile phones were exempt as they were not considered to be devices that produced exposures in excess of the SAR limits. At that time Health Canada asked the Royal Society of Canada to review the Safety Code 6 with respect to its adequacy in protecting the public from potential biological and health effects of radio frequency fields from mobile phones. A panel of eight scientists was asked to answer a number of questions including the following: "Is there evidence that such non-thermal effects if any, could be greater for children or other population subgroups?" The panel decided that generally "while a number of biological effects have been observed (on cells or animals) to non-thermal/athermal levels of exposure to RF fields, these biological effects are not known to cause adverse health effects in humans or animals," and in response to the question regarding specific risks to children they concluded that "there has been little study regarding the greater susceptibility of subpopulations to RF field exposure." The Royal Society of Canada conducted a review of the panel's findings in 2004 and concluded that "all of the authoritative reviews completed within the last two years have concluded that there is no clear evidence of adverse health effects associated with RF fields" (British Columbia Centre of Disease Control, 2009).

SAR uses thermal rate of change as the principal determinant of health effects. According to Safety Code 6 mobile phones in Canada are deemed to be safe as long as they do not cause the thermal effects in excess of the limits. However, it is very difficult for member of the general public to determine whether the conditions of day-to-day use of a mobile phone is within Safety Code 6 limits (e.g., factors such as power, distance from base stations, attenuation, antenna technology, etc. should be factored into the analysis). The Institute of Environmental Medicine at Karolinska Institute in Sweden found that 10 or more years of mobile phone use increased the risk of acoustic neuroma and that the risk increase was confined to the side of the head where the phone was usually held. However a parallel study about mobile phone use and the risk of brain tumors did not indicate any increased risk of brain tumors in relation to mobile phone use.

As the only code governing mobile phone EMF in Canada, Safety Code 6 has been challenged by numerous groups as outdated and insufficiently grounded in detailed research about use of mobile phones in Canada. For responses to questions posed to Health Canada regarding EMF and mobile phones Safety Code 6 remains the only source.

We question the utility and reliance on Safety Code 6 as Health Canada's guide on EMF. This code is not written for the general public and is not based on research of how Canadians and particularly young Canadians use mobile phones. On the provincial and municipal levels there are perspectives that conflict with the federal stance on mobile phone use and EMF. For example, in 1999 the Ontario Chief Medical Officer of Health, the late Dr. Sheila Basrur, commissioned independent research on mobile phones and EMF health effects and issued a memo in November of 1999 to the City Council based on the results that recommended the following for the City of Toronto:

1. Practice prudent avoidance—site towers at least 200 meters from schools and day-cares.
2. Reduce radiation emissions by 100 times below current limits.
3. Notify residents of new or changed installations, allowing citizens to be an integral part of the process.

Dr. Basrur further recommended guidelines for exposure to microwave radiation as emitted from cellular phones and antennae, 100 times more stringent than those that currently exist. She remarked that Health Canada's claim that Safety Code 6 is "one of the most stringent in the world" is simply not true. Switzerland's emission standard is 90 percent lower than ours, Italy's is 85 percent lower, the province of Salzburg, Austria is 98.5 percent lower, and six municipalities in Australia have standards 10,000 lower than our "stringent" standards (: http://emr. bc.ca/Articles/TwoCities.html). To the Canadian Press Dr. Basrur is reported to have said, "It disturbs me that kids are the marketing target for devices that are dressed up to look as innocuous and friendly as possible, and yet may have longer-term health implications attached to them that we're not fully aware of.... It falls on government and industry to provide this information in a readily accessible, easily understood fashion so you don't need a post-doctorate degree in radiation physics to realize that the jury is out."

In summary, while there is some consensus that thermal effects from mobile phone EMF are thus far unsubstantiated, there are unresolved concerns about biological effects. Among member countries of the European Union this uncertainty expressed by the scientific community has lead to a spirit of precaution with individual country health institutions recommending reductions in unnecessary mobile phone EMF exposure while research continues, especially in the case of mobile phone use by children and young people. Health Canada has not adopted a similar approach and has taken the opposite stance.

Based on the same so-called lack of evidence driving European precaution, Health Canada has opted not to issue any substantial cautionary statements regarding mobile phone use either for the general public or specifically for Canadian youth. This apparent requirement by Health Canada for more stringent scientific evidence of detrimental health effects has not necessarily been required when issuing social policy recommendations in other cases. In the following section we describe Health Canada's policy approach regarding another recent controversial issue, the use of bis phenol A, and we demonstrate an apparent contradiction in the burden-of-proof of negative health effects as a requirement to initiate social policy for EMF.

Comparing BPA and EMF Policy Responses in Canada

To address the assertion that one reason for the weak policy response of Health Canada regarding mobile phones and EMF is due to a lack of conclusive scientific evidence we compare and contrast Health Canada's response to another controversial issue. In October 2008 Canada became the first country in the world to take regulatory action and ban the use of a widely used chemical, bis phenol A (BPA), in infant food containers. In a statement to the public, the then Minister of Health Tony Clement stated that, "we have immediately taken action on bis phenol A because we believe it is our responsibility to ensure that families, Canadians and our environment are not exposed to a potentially harmful chemical." This policy action came about because, according to Mr. Clement, Health

Canada issued a challenge to industry in 2007 to provide information on how they manage negative health effects of bis phenol A.

The results from research into the health implications of BPA were inconclusive. According to Health Canada, "scientists concluded that BPA exposure to newborns and infants is below the levels that cause effects." In addition, the broader international scientific community is far from unified, or in truth, even prioritizing possible BPA concerns as a pressing concern. WHO has shown little interest in BPA, and the U.S. National Institute of Environment Health Sciences identified some concern, but did not issue any regulatory action.

Despite the lack of conclusive scientific evidence of negative health effects, the Government of Canada banned the use of BPA due to the uncertainty raised in some studies related to the potential effects of low levels of BPA. Mr. Clement stated that "the Government of Canada is taking action to enhance the protection of infants and young children." Moreover, the Canadian Government allocated a further $1.7 million over 3 years to fund research projects on BPA.

In contrast, as of June 2009 Canada is one of the only developed countries to take no policy action regarding EMF. A quick international survey of EMF related policies reveals that other G8 countries have already taken precautionary measures. France and Germany have both instated strong recommendations with respect to safe use of mobile technology for youth. Citing Safety Code 6, and statements on inconclusive scientific evidence by WHO, Health Canada has determined that there is insufficient evidence that EMF radiation from mobile phones is harmful. On June 17th, 2009 Tony Clement—now in the capacity of the Minister of Industry approved a controversial building permit for a mobile phone tower in Charlottetown, P.E.I. In doing so he claimed that "Health Canada reviews [EMF] concerns from a health and safety point of view. Health Canada did its job and reviewed EMF for health and safety concerns. Those concerns were not valid. On that basis, that is why these are left to the national level to ensure we have the network capacity that is necessary for health and safety reasons, for instance 911 calls and things like that...."

In fact this statement is misleading. The WHO and the international scientific community have clearly placed research on the possible effects of exposure to EMF as a priority, with strong discussion and contention on both sides of the debate. WHO maintains a vast website with extensive literature available from researchers across the globe, including peer reviewed articles, and detailed database information regarding international safety standards. Furthermore, they are currently co-coordinating the International EMF project. According to WHO, they "established the International EMF Project in 1996 to assess the scientific evidence of possible health effects of EMF in the frequency range from 0 to 300 GHz. The EMF Project encourages focused research to fill important gaps in knowledge and to facilitate the development of internationally acceptable standards limiting EMF exposure."

Health Canada, however, after referencing participation in the International EMF Project asserts "there are no Canadian Government guidelines for exposure to EMFs regarding mobile phones. Health Canada does not consider guidelines necessary because the scientific evidence is not strong enough to conclude that typical exposures cause health problems." What is particularly problematic about this statement is the assertion that typical exposure does not cause problems. Health Canada makes the identical statement about BPA, but enacted regulatory action, with specific reference to possible health effects on infants and young children. With respect to EMF, Health Canada, in an effort to support their decision of non-action on EMF provides links to various studies for further information, including the 2002 NIEHS study which states "Since 1999, several other assessments have been completed that support an association between childhood leukemia and exposure to power-frequency EMF."

While health Canada is not only selective about how it uses similar information, and effectively tries to shut the EMF debate down, Minister Clement took it one step further this week when he overturned The City of Charlottetown's Council decision to refuse permission for Rogers Cable to build a cell phone tower on a particular piece of property. Permission was denied, in part based on concerns about EMF exposure. Minister Clement overturned the Council decision on June 17, 2009 from the floor of the House of Commons.

One common link between these two issues is the emphasis of the possible effects on children and youth, a topic that has recently become more prominent. There are several prominent studies awaiting publication, including the highly anticipated International Association for Cancer Research (IARC) Interphone study, and ongoing contributions from the International EMF project. With respect to media interest, CBC's Marketplace aired an episode questioning why Canada has not made any recommendations regarding youth and cell phone use when other G8 countries, in particular France, have made strong recommendations (Jan. 2009). During the course of the program, Dr. Elizabeth Cardis, co-coordinator of the unreleased Interphone study identified

EMF health hazards for youth as a serious concern requiring much more investigation and caution, and made preliminary usage and practice recommendations. In the U.S., Larry King Live dedicated an entire show to the potential hazards of EMF exposure and youth. Finally, as we'll discuss, academic research examining the possible links between youth and EMF exposure through mobile phone use, of peer-reviewed publishing standard, is becoming more common. Frustratingly, Minister Clement would be fully aware of the impending release of the Interphone study as it is a highly prominent study, and is in fact being coordinated by a Canadian researcher. Yet, he continues to dismiss any possibility of serious concern for EMF and health risks, despite regulatory action being taken by the European Union based on the same observations made by Dr. Cardis. These recommendations will be considered in more detail in the next section.

Mobile Phone Practices among Canadian Youth

Another documented (Burgess 2003) reason for the lack of policy response from Health Canada regarding EMF and mobile phones is that mobile phone use in Canada is not as great as in other countries where more proactive policies are being implemented. This premise, however, is not grounded in the empirical data on mobile phone use in Canada today.

Canadians and the Mobile Phone

Within the last 10 years mobile phones have become integral to the lives Canadians and essential to the social and economic composition of Canada. According to the Canadian Wireless Telecommunications Association (CWTA), more than 25,000 Canadians work in or for the mobile industry in a country of approximately 33 million people. The industry generated more than $12 billion Canadian in 2007. At the end of December 2008, Canadian wireless phone subscribers numbered 21.5 million, representing a national wireless penetration rate of 67 percent. CWTA research estimates wireless penetration in major urban centers has exceeded 70 percent, with some greater metropolitan areas approaching the 80 percent mark.

Canada has an atypical distribution of what is considered to be relatively small population across a large and in many places geographically remote land mass. Despite this two-thirds of Canadian households have access to a mobile phone. Canadians reportedly sent an average of 77 million text messages per day in 2007 and wireless revenues in Canada totaled \$12.5 billion in the same year. Half of all phone connections in Canada are now wireless (CWTA, http://www.cwta.ca/CWTASite/english/index.html).

Youth Practices around the Mobile Phone

In all countries/cities where EMF concerns have lead to social policy, regulation or guidelines children and young people are deemed to be a group for whom the consequences of exposure are most concerning. Health Canada has not distinguished younger mobile phone users from the wider population. We argue that two factors account for this; first, there is a paucity of information on mobile phone use among Canadian young people; and second, policymakers do not have the analytical tools to effectively deal with young people who represent not quite a dependent (as in the case of infants an bis phenol A) but not quite consumer citizens either. In the following section we address the first factor—examining how young Canadians are using mobile phones. Further on, we'll investigate the second factor using policy frameworks from Schneider & Ingram (1993) and Hacker (2004, 2005).

We focus on youth in this chapter since research has shown that the biological physiology of the brains of younger children and youth permit deeper penetration of EMF radiation than the brains of adults. In addition, children and youth are traditionally viewed as policy targets, with particular emphasis on protection. Most recently, the European Parliament overwhelmingly voted to adopt the Ries Report, with the following resolution: "whereas most European citizens, especially young people aged from 10 to 20, use a mobile phone, an object serving a practical purpose and as a fashion accessory, and whereas there are continuing uncertainties about the possible health risks, particularly to young people whose brains are still developing."

A German cancer researcher, Joachim Schz (2005), states that the main difference concerning the use of mobile phones between today's children and adults is the longer lifetime exposure of children when they grow older, due to starting to use mobiles at an early age. Schz notes that the prevalence of mobile phone users among adolescents is over 90 percent in some countries. For example, in a German study, 6 percent of 9-10 years old children used a mobile phone for making calls daily; 35 percent owned their own MP.

The Public Health Agency of Canada released the Healthy Settings for Young People in Canada report. In the report, they highlight the crucial link between practice and policy, demonstrating that Canadian policymakers are well aware of the importance of linking youth practices and policy recommendations in a unified manner:

> The Health Behaviour in School Aged Children Study (HBSC) is a valuable tool for monitoring changes in health behaviors and important health outcomes among young Canadians. The broad range of health-related issues addressed in the survey reflects the complicated lives of young people. The report identifies the importance of the links between physical, social, and emotional states, as well as between contexts, behaviors, and outcomes. By examining family and peer relationships, the school setting, and socio-economic status, we can gain insight into the strong impact certain settings and conditions have on risk-taking behaviors and health outcomes.

Apart from general information from CWTA and Statistics Canada on use of mobile phones within the adult population, very little research has been conducted on youth practices. There is general folk-framing that Canadian children and teens are among the most intense users of mobile phones and are doing so in increasing numbers. In this vein Toronto-based Solutions Research Group made a forecast in 2006 that within a few years one in five children aged 8 to 11 would have a mobile phone. These statistics are unconfirmed as Statistics Canada, the body that conducts government-funded research, has to date not collected this kind of data about young mobile phone users.

However, the Canadian Broadcasting Corporation (CBC) surveyed 1,084 young people aged 9-13 in schools from Calgary, Toronto, Brampton and Waterloo in November and December of 2008 and found that 40 percent owned their own mobile phone.[1] Although this was not a rigorous study, the results resonate well with what we see and hear in everyday life in urban cities.

In a recent survey of mobile phone practices of young people in Toronto (McEwen, 2010),[2] over 82 percent of 173 young people 17-30 years of age professed to using their mobile phone every day, and spent an average of 300 minutes per month on the mobile phone. Furthermore, although about three-quarters of them had access to a landline in their residences, less than one-quarter of them used it.

Over 80 percent of participants surveyed reported that they do not turn their mobile phones off; in fact, when asked where they put their mobile phones once they have gone to bed, many of them report that the device is on and within arm's length—either under the pillow, on a bedside table or on the floor next to the bed. Even if they are asleep, the potential to be woken up for a call or text message is acceptable, and in the case of romantic exchanges, even desired. Caron and Caronia (2007) call Canadian youth between the ages of 17 and 22 the "on generation," and state that "young people under twenty are the first generation to have known from infancy such a wide-ranging media landscape—new technologies have always been part of their framework of experience and social relations" (p. 200). More than half of the interview participants take this even further: they view their mobile phone as more than part of their media landscape and they feel that their mobile phone is a part of them. In this sense, they believe that if they are always "on"—that is, ready and able to communicate at all times, their phone should be as well. Consider the following excerpt from an interview:

Interview—Gerri, 18, from a suburb of Toronto.

Gerri: When I go to bed, the mobile phone is right beside me on the mattress. It wakes me up about every half hour.

Interviewer: No straight night's sleep?

Gerri: No.

Interviewer: So do you leave the ringer on or off?

Gerri: It's on vibrate because I wouldn't wake up with the ringer but vibrate will shake my whole mattress. I'm pretty protective of it. If somebody takes [my mobile phone] away from me and tries to go through it, I feel it's like my diary because I have text messages from everybody that they can read, and it really does tell a lot about me and my life. And when my friends goof off and start kicking it around—for a joke, 'cause it's old I get offended! It's like they're hurting me! So I do feel that it is a part of me, like it's constantly with me.

This participant was representative of over half of the participants interviewed (n = 12) who described the mobile phone as "a part of them." The mobile phone is integrated into the everyday lives of Canadian young people to the extent that it has become a taken-for-granted part of socialization.

An understanding of Canadians, and particularly the heavy mobile phone using young Canadians, is an important and missing factor in policymakers' consideration of how to handle EMF issues.

Seven years worth of internal Health Canada documents, obtained through access to information requests, reveal concerns about cell phone frequencies and potential—but unproven—links to "childhood leukemia, brain and other cancers of the head and neck, memory problems, stress and migraine/neurological ailment."

There are studies that indicate that children are at the highest risk from EMF exposures (Kheifets, Repacholi, Saunders & van Deventer 2005; Olsen, Nielsen & Schulgen 1993). Based on these and similar findings WHO recommends that epidemiological research on EMF effects on young people be given the highest priority. WHO states that

> [S]ince many children are heavy mobile phone users and will continue to be in the future, they represent a unique population. The type of mobile use among children (e.g., text messaging), their potential biologic vulnerability and longer lifetime exposure make such a study desirable. Cognitive effects and other general health outcomes have been anecdotally reported in mobile phone users. They can be assessed in a prospective cohort study of children. A separate study of children was found necessary, as it is not possible to just extend the age range of a cohort study of adults because the outcomes have to be assessed by different methods in children and adults, and children's exposure probably differs from that of adults (more use of pay-as-you-go SIM-cards, more frequent change of phones and operator).[3]

In contrast, young people do not represent a unique category of users in Safety Code 6. We suggest that the lack of empirical data on youth mobile phone practices translates into lower level of concern about this group by Health Canada. While practices are clearly considered in the development of EMF policy-decisions other countries and WHO, it is not the case in Canada. Instead, Canadians who visit the Health Canada's website are simply instructed to decide for themselves whether they can live with the "possibility of an unknown risk from cell phone use." Health Canada recommends using a headset if you have concerns about EMF. This is a perfect example of the way policy recommendations from the regulator is out-of-touch with youth practices. In McEwen's study more than 60 percent of those interviewed indicated that they had some concerns about EMF. One participant summed it up as follows: "[A]pparently, if you talk on the [mobile] phone too much you get brain tumours. And then, a lot of times if I talk on the phone for long time, like an hour at a time, and your head just goes warm and you wonder "Oh my God am I getting a tumour?" I have a headset, and a couple times I fished it out but not really—I use my phone everywhere and it's not convenient to carry around the headset—you know?"

In the next section we analyze the manner in which different policy institutions globally have articulated their positions regarding EMF, and conduct a brief document analysis to identify the ways that Canadian policymakers have or have not aligned with other policymakers. This is important to (a) emphasize the point that Canadian social policy regarding EMF is inconsistent with those in similar countries, and (b) contest statements by Health Canada that Canada's approach on EMF is based-on recommendations from the WHO.

Social Policy and EMF

For any researcher attempting to wade into the maze of international policy and EMF, it is quickly apparent that there are many, many people, industries, and policy actors to sort through. Furthermore, there are wide ranging discrepancies between countries with respect to EMF exposure standards, the acceptability of different research, and the broad ministries and departments responsible for developing policy.

In Canada, it is very difficult to determine who exactly is responsible for determining the legitimacy of scientific evidence regarding EMF exposure. Health Canada, Industry Canada, the CRTC, the Telecommunications Industry, academics, public interest groups, and politicians in general all seem to have input.

Globally, the picture becomes even more muddied. WHO appears to be the central organizing body, the European Parliament, IARC, the International Commission on Non-Ionizing Radiation Protection (ICNIRP), the EU, and various independent bodies, National Health Organizations, Telecommunication Industry participants and public interest groups are all involved.

Generally, while the EMF policy community appears to revolve around WHO, we suggest that WHO's authority is more symbolic. As a small example of this symbolic power, we analyze the contents of documents from the vast library WHO has created on EMF. As a part of their International EMF Project, WHO created three comprehensive documents for drafting policy. The first, Establishing a Dialogue on Risks From Electromagnetic Fields, provides step by step guidance for policymakers to begin a public debate about EMF and health concerns. In particular, they identify three key discussion areas: risk assessment, risk perception, and risk management. Despite the fact that the document was created for policymakers, and in general, provides broad guidelines for policymakers to get their message across rather than engage in a genuine debate, Canadian Policy-makers have not, to our knowledge, engaged in any significant degree of policy debate. In 2006, WHO released two more documents: Framework for Identifying Health-Based EMF Standards (June 2006), and Model Legislation for Electromagnetic Fields Protection (2006). In the first, WHO draws attention to the need to harmonize international standards, and in particular "develop science-based exposure limits that will protect the health of the public and workers from EMF" (p. 5). However, it can be suggested that when it comes to EMF policy in Canada, this recommended framework has not been considered in any serious way. For example, at one point, it is recommended that "as new scientific information becomes available, standards should be updated. Therefore, a mechanism for periodic scientific evaluation by an appointed council should be set up that would issue, where necessary, amendments to the standard." It is fairly clear that Canada is not currently interested in amending our current policy on general EMF exposure, let alone creating a new policy specifically targeting youth with respect to mobile information practices.

Finally, WHO released model legislation for regulating EMF exposure, for any country to use and adapt as necessary. To our knowledge, this resource has not been used by policymakers in Canada, as they instead continue to steadfastly adhere to the use of Safety Code 6 as the only regulatory guideline for EMF exposure.

One thing that is lacking from most of the policies, recommendations, and regulatory acts, is an acknowledgment of an understanding of the information practices of youth and their mobile devices. It can be suggested that the current policy in Canada, drafted in 2000 and updated in 2004, does not base any policy recommendations on a complete understanding of the true nature of youth exposure to EMF via mobile phone. In part, this is due to a total lack of survey data on this topic by Statistics Canada. However, on another level, it could be due to an ideological shift in the way government considers the needs of youth in this policy area. This notion will be explored further in the Discussion section of this chapter. Policy based in 2001 cannot possibly address the current practices of Canadian youth and their mobile phones.

Unlikely Reasons for Canadian Policymakers' Silence

As we move into an in-depth discussion examining the possible factors contributing to a lack of debate in Canada on this topic, we'd like to first take up a brief overview of some of the less complicated factors.

To begin, it is useful to rule-out some of the possible inferences from the chapter so far. It should be made clear that, despite the complete lack of debate and policy action, it is not being suggested that any policy actors in Canada are maliciously attempting to manipulate the policy process, nor is any policy-actor being deliberately neglectful. Along those lines, it is not being suggested that Health Canada and Industry Canada are willfully beholden to the Telecommunications Industry in Canada.

However, it should be suggested that the political terrain is muddy. In particular, it appears to be over-saturated with complicated, difficult to decipher and compare/contrast research. Moreover, the policy actors are woefully under informed about the true information practices of Canadian youth, given the ubiquity of mobile communication. Furthermore, the indeterminate nature of the terms youth, child, and teen can be shown to contribute to the opacity of the information practices of these groups in Canada. Not only is there a lack of understanding about these practices, but there is a lack of understanding about who is actually using what, and at what age.

In the previous sections we have in turn questioned the validity of three other explanations for inaction. We believe that (a) a lack of conclusive, scientific evidence of harm, (b) general perception among the public that Canadian regulators follow other countries but do not lead, and (c) low-adoption rates of mobile phones, are (in some cases false) explanations that may influence but not cause inaction regarding mobile phone EMF.

In the following sections we offer an alternate theory for why social policy on mobile phone EMF has been slow to develop in Canada.

Discussion

In this section we consider the issue of policy process more deeply. Drawing from Social Construction theory, and Hacker's concept of Policy Drift, it can be suggested that Canadian youth are viewed as a group in transition; or to put it more specifically, they are transition from a target population to a market population. Schneider and Ingram (2008), in discussing social constructions of target populations in public policy arenas note that their framework "contend[s] that the policymaking process often involves the creation of 'target populations' and then uses or creates images of the target groups that will justify the allocation of benefits or burdens to that group" (p. 191).

Traditionally, in Schneider and Ingram's framework, policy targets viewed as dependents are described as having a "positive construction as loving, sweet, blameless, helpless, 'good' people such as mothers and children, but they exercise almost no political power" (p. 193). Political power is defined in many ways, but for our purposes, it is considered along the lines of "level of participation in traditional activities such as voting, giving, campaigning, and so on is very low" (p. 195). Central to the social construction framework is its recognition of the "interaction between political power and social constructions contending that public policymakers and the policy-making process itself manipulate constructions, bringing some to the forefront and pushing others into the background" (p. 195).

Thus, with respect to the protective policies of bis phenol A, Canadian youth are conceived as dependents, and have correspondingly low political resources. They are not likely, as young children and infants, to be voting or raising money for political parties. The policy appeals to mothers as well, who are viewed as generally falling between dependents and advantaged with respect to power, in that the policy creates a mandate for industries to innovate to produce safer products for infants.

Conversely, with respect to EMF policies, it can be suggested that Canadian youth are conceived as an advantaged group—a group that is "assumed to be worthy" (p. 193), and are often associated with wealth. While it is true that Canadian youth are not traditionally associated with wealth, that conception becomes skewed as a clearer understanding of their practices with mobile devices is revealed. They become viewed as a group with significant disposable income, often spent on the latest mobile phone released on the market. They have also shown themselves to be extremely savvy in their information practices with respect to income management—shifting their usage practices (texting vs. talking) to accommodate their income. Thus, the lack of regulatory policy regarding mobile phone use would be viewed in a positive light by young Canadians.

Schneider and Ingram (1993) continue to consider the implications of citizenship for targeted populations, suggesting that "the agendas, tools, and rationales of policy impart messages to target populations that inform of their status as citizens and how they and people like themselves are likely to be treated by government" (p. 340). In doing so, it can facilitate the transition from dependent policy target to advantaged policy target with a new conception of citizenship rights, one that is "internalized [and] influences their orientations toward government and their participation [...] telling them whether they are viewed as 'clients' by governments and bureaucracies or whether they are treated as objects" (p. 340).

The current policy situation also appeals to parents. In particular, surveys have revealed that one of the main reasons youth receive mobile phones in the first place is based on the desire of parents to maintain a semblance of protective security for their children. Mobile phone ownership has been shown to provide this sense of security to parents.

Thus, it can be suggested that the Canadian Government is taking advantage of this dual conception of Canadian Youth to create a policy environment that is advantageous to their oft-stated desire to continue promoting the commercialization of the telecommunications industry.

In effect, youth are portrayed politically as needing protection, but are, in actuality, perceived politically as consumers/clients. This contradiction fits with Schneider and Ingram's assertion that groups are fluid—changing and shifting as social constructions are "manipulated and used by public officials, the media, and the groups themselves" (p. 342).

Furthering their analysis, Schneider and Ingram consider the different types of policy design that are created for different types of targets. Of particular interest for our purposes is their observation that: Advantaged targets—those both powerful and constructed as "deserving"—typically enjoy the professionalized and respectful treatment of federal career employees, whereas dependents are far more likely to be served by state or local administrators who have considerable discretion in whom they do or do not treat well (2009, p. 205).

As noted earlier, the Toronto Board of Health recently adopted recommendations with respect to the mobile phone use practices of children and youth, suggesting that youth are firmly targeted as dependents. While at the federal level, as has been made abundantly clear throughout this chapter, there are no guidelines or recommendations with respect to mobile phone use and Canadian youth, perhaps inferring that the federal government does in fact view youth as a dependent construct. But, it is too facile to simply view the policy execution in such a two dimensional manner. As we have demonstrated with Schneider and Ingram, we can suggest that youth are a group in transition. But how is this transition accomplished. We posit that it is, in actual fact, the lack of a debate in and of itself that is facilitating the transition of the social construction of youth as a dependent group to an advantaged group. And furthermore, the lack of debate contributes to accomplishing the transition in a manner that appears to be as natural and fitting as the execution of the policy process leading to the regulation of BPA.

Drawing on Hacker's theory of policy drift (2004, 2005), we can begin to theorize about the role of a "non-debate" in transitioning youth and social construction. Hacker conceives of policy drift in his analysis of welfare state retrenchment in the U.S., but his discussion is relevant here too. He argues that policy drift is most simply defined as "changes in the operation or effect of policies that occur without significant changes in those policies structures" (p. 246). In particular, he claims:

> The major cause of drift in the social welfare field is a shift in the social context of policies, such as the rise of new or newly intensified social risks with which existing policies are poorly equipped to grapple. The hallmark of change of this sort is that it occurs largely outside the immediate control of policymakers, thus appearing natural or inadvertent. The question for policymakers becomes whether and

how to respond to the growing gaps between the original aims of a policy and the new realities that shifting social conditions have fostered (Hacker, 2004, p. 246).

It is perhaps best to discuss this notion more concretely to help tease out the connections. As highlighted earlier, Minister Clement overturned The City of Charlottetown's decision to deny a building permit for a cell phone tower. Minister Clement argued that he has a responsibility to facilitate the construction of communication networks across Canada, in part to ensure sufficient coverage for mobile users should they need to call 911. However, as McEwen's (2010) research shows, the social practices associated with youth and mobile phones have shifted considerably over the past several years, with usage increases in both talk and text use, moving beyond simply using phones for emergencies such as 911.

Possible health risks have been identified both in the global health community and the academic community; with these risks increasing as the social practices around mobile phone use continue to shift towards greater and more frequent use of mobile devices across many interface platforms.

To peel back the layers of this policy conundrum further, the Health Canada policy on EMF exposure, based in particular on Safety Code 6 guidelines, was implemented when Canadian youth were clearly constructed in policy as dependents—as requiring protection. However, the social practices of youth with their mobile devices have far outgrown the expectations that were in place when Safety Code 6 was enacted. So, as Hacker observes, the question facing Canadian policymakers was whether to, and/or how to, respond to the growing gaps between the Health Canada policy, and the actual usage practices observed. We posit that policymakers in Canada have chosen not to respond—to let the policy drift so to speak, in order to facilitate the transition of youth as a protected, dependent construction into an advantaged, desired clientele with discretionary spending potential. Hacker argues: "In an environment of new or worsening social risks, opponents of expanded state responsibility do not have to enact major policy reforms to move policy toward their favored ends. Merely by delegitimizing and blocking compensatory interventions designed to ameliorate intensified risks, they can gradually transform the orientation of the existing program" (Hacker, 2005, p. 46).

While Hacker is speaking specifically about welfare retrenchment and institutional change, his argument provides relevant guidance for our discussion. The federal government delegitimizes any debate around the possible health effects of EMF and youth, and effectively shuts down the discussion. Therefore, youth are transitioned out of the domain of Health Canada with respect to EMF exposure, and are quietly socially reconstructed as citizens participating in the continued promotion and expansion of Canada's Telecommunications Industry.

(In conversation with an MP familiar with EMF policymaking processes in Canada, our conjectures were partially/preliminarily confirmed. He felt there was still confusion amongst policy makers about the possible dangers of EMF, and indicated a continued reliance the relevance and applicability of Safety Code 6. Furthermore, the issue of usage practices had not been, in his opinion, considered at all by policy makers. Thus, in reference to our conjecture, there are clear indications of an over reliance on Safety Code 6 amongst policymakers and poor understanding of the implications of the newly emerging mobile practices of youth in Canada.)

This discussion is further illuminated by comments made by John Wilkinson, Minister of Research and Innovation for the province of Ontario, and Premier Dalton McGuinty at the recent Canada 3.0 Conference. Wilkinson claimed that the role of government is not to interfere, but to be a catalyst—that the marketplace should be free to innovate and determine our needs, and most importantly that "we should let science solve the problem, and not let politics and political science get in the way"; that we should facilitate action by not acting. Premier McGuinty confirmed the critical position of the ICT industry in the Canadian economy with his assertion that "technology is a race to get it into the hands of consumers" (June 9, 2009). These statements open the door for a further debate about the instrumental vs. structural roles of governments and their relationships with regulated industries.

Conclusion

The primary objective of this chapter was to review and analyze the factors that may or may not contribute to how Canadian policymakers consider action with respect to possible negative health effects from mobile phone EMF. In particular, we questioned the lack of precautionary guidelines in the case of young people given their high usage of mobile phones in line with similar countries.

Using BPA as an example we questioned the validity of other widely touted explanations for inaction. We believe that (a) a lack of conclusive, scientific evidence of harm, (b) general perception among the public that Canadian regulators follow other countries but do not lead, and

(c) low-adoption rates of mobile phones, are explanations that have not been the reason for inaction regarding mobile phone EMF.

Instead using empirical data on the mobile phone practices of young people, and discourse analysis we conjecture that an explosive adoption of mobiles by Canadian young people has caught policymakers off-guard and that the lack of visibility of youth mobile phone practice within the policy community, and a complicated policy environment surrounding this issue have culminated in silence.

This chapter represents a first step in investigating Canadian social policy and the mobile phone. While we have focused on Health Canada and mobile phone use in Toronto, further research on the discourses of other policy institutions in Canada would enhance the analysis provided in this chapter. For future research, it would be useful to draw on other theories, such as Schoen and Rein's theory of framing and intractable policy controversies to illuminate the policy processes at play. In addition, while we have indicated that the policy environment surrounding mobile phone EMF is complex, a mapping of the policy environment would be a positive contribution to the literature, and would be a useful tool for both policymakers and activist groups alike.

We believe that research that endeavors to set-aside the specifics of whether or not mobile phone EMF is true or false in order to understand the implications of this issue from a social policy perspective can highlight broader issues, such as the consequence of limited data on technology practices on policy. We are certain that in the future other cases will arise that in the heat of the controversy obscure underlying gaps in the policy process. We therefore encourage other analyses, such as this one, that attempt to highlight the importance of understanding information practices in the development of social policy.

Methodological Note

McEwen (2010) analyzed the everyday use of the mobile phone by first-year undergraduates in Toronto over the 2007-2008 academic year, gathering survey and interview data at three points. Participants were English-speaking, first-year, undergraduate students of any program of study, full-time or part-time, between the ages of 17 and 22, who were already mobile phone users at the downtown Toronto campuses of the University of Toronto and Ryerson University. In the first semester the first-wave of the online survey was issued in November (Nov 2007; n = 173), eight weeks after the semester started. The second-wave of the survey was issued to the same participants in the following March

(Mar 2008; n = 153). For completing the questionnaire both times participants were entered into a raffle with a 1-10 chance to win high-tech devices. This drew a 95 percent response rate. The second part of the study involved in-depth interviews of 20 of the survey participants during July and August. Sociograms, or visualizations of their social networks, were developed during the interview to examine social network changes over their first year. McEwen employed techniques from Social Network Analysis to design both the online survey and interview instruments. The data collected included the following statistics and qualitative statements: number of mobile phone calls made and received; number of text messages made and received; the importance of the mobile phone compared with other communication devices; the types of people communicated with using a mobile phone; the general content of mobile phone calls; the amount of time spent on mobile phones; the relationship time of the persons contact by mobile phones (i.e., very close friends, somewhat close, acquaintances, strangers); when the phone is on or off; what the phone looks like physically (use of personalization); etc.

Notes

1   Source: CBC, http://www.cbc.ca/marketplace/2009/generation cellphone/survey. html, last viewed June 23, 2009.
2   See endnote for a methodological note on the study.
3   Source: WHO website, http://www.who.int/peh-emf/research/children/en/index4. html, last viewed June 16, 2009.

References

British Columbia Centre of Disease Control. Retrieved June 22, 2009 from http://www. bccdc.org/content.php?item=55.
Burgess, A. (2003). *Cellular Phones, Public Fears, and a Culture of Precaution*, Cambridge University Press, London, UK.
Caron, A.H., & Caronia, L. (2007). Moving Cultures: Mobile Communication in Everyday life, McGill-Queen's University Press, Montreal, Canada.
Clement, Tony. (June 17, 2009). 40th Parliament, 2nd Session: Edited Hanard No. 77; Wednesday, June 17. 2009. House of Commons, Canada.
European Union. (February 23, 2009). Report on Health Concerns Associated with Electromagnetic Fields: A6-0089/2009. http://www.europarl.europa. eu/sides/getDoc. do?type=REPORT&reference=A6-2009-0089&language= EN.
Hacker, Joseph. (2004). Privatizing Risk without Privatizing the Welfare State: The Hidden Politics of Social Policy Retrenchment in the United States. *American Political Science Review*. Vol. 98, No. 2. pp. 243-260.
Hacker, Joseph. (2005). "Policy Drift: The Hidden Politics of US Welfare State Retrenchment." In *Beyond Continuity: Institutional Change in Advanced Political Economies*. Oxford University Press: UK. pp. 40-82.
Health Canada. (2004). Electric and Magnetic Fields at Extremely Low Frequencies. http:// www.hc-sc.gc.ca/hl-vs/iyh-vsv/environ/magnet-eng.php. Retrieved June 20, 2009.

Health Canada. (2008). Government of Canada protects family with Bisphenol A Regulations. http://www.hc-sc.gc.ca/ahc-asc/media/nr-cp/ 2008/2008 167-eng.php. Retrieved June 20, 2009.

Health Canada. (2008). Ministers Remarks on Bisphenol A. http://www.hc-sc.gc.ca/ ahcasc/minist/speeches-discours/2008 04 18-eng.php. Retrieved June 20, 2009.

Jensh, R. (1997). "Behavioral teratologic studies using microwave radiation: Is there an increased risk from exposure to cellular phones and microwave ovens?," *Reproductive Toxicology*, Volume 11, Issue 4, July-August 1997, pp. 601-611.

Kheifets, L., Repacholi, M., Saunders, R. & van Deventer, E. (2005). "The Sensitivity of Children to Electromagnetic Fields", PEDIATRICS Vol. 116 No. 2, pp. 303-313.

Linder, S. (1995). "Contending discourses in the electric and magnetic fields controversy: The social construction of EMF risk as a public problem," *Policy Sciences*, Volume 28, Number 2.

Matthes, R. (1992). "Radiation emission from microwave ovens," Journal of Radiological Protection, Vol. 12, pp. 167-172.

McEwen, R. (2010). "A world more intimate: exploring the use of mobile phones in maintaining and extending social networks," Ph.D. dissertation, Faculty of Information, University of Toronto, Canada.

National Institute of Environmental Health Sciences and National Institute of Health. (2002). EMF: Electric and Magnetic Fields Associated with the Use of Electric Power. NIEHS/DOE EMF RAPIDProgram. http://www.niehs.nih.gov/health/topics/ agents/emf/ Retrieved June 20, 2009.

Olsen, J., Nielsen, A. & Schulgen, G. (1993). "Residence near high voltage facilities and risk of cancer in children," BMJ, Vol. 307, pp. 891-895.

Public Health Agency of Canada. (2008). Healthy Settings for Young People in Canada. Minister of Health: Ottawa, Ontario.

Schneider, Anne & Ingram, Helen. (1993). "Social Construction of Target Populations: Implications for Politics and Policy." *American Political Science Review*. Vol. 87, No. 2, pp. 334-347.

Schneider, Anne & Ingram, Helen. (2008). "Social Constructions in the Study of Public Policy." in *Handbook of Constructionist Research*. Holstein, J. & Gubrium, J. (eds). The Guilford Press. New York: New York, pp. 189-211.

ScheiderWorld Health Organization. (2006). *Framework for Developing Health-Based EMF Standards*. WHO Press: Geneva, Switzerland.

10

Mobile Communication Policies in the Workplace: The Case of U.S. State Governments

Craig R. Scott, Hyunsook Youn,
and Gillian Bonanno

Mobile phones are one of the fastest growing technologies over the last decade, averaging 24 percent annual growth (Wray, 2008). Subscribers across the world passed 2.17 billion in late 2005 and are expected to reach 4 billion by 2010 (Kelly & Biggs, 2007). Although the increasing use of mobile phones and other portable devices (e.g., personal digital assistants (PDAs), laptops, etc.) has brought desired changes in work and society (Wei & Leung, 1999), these devices can also be used in less intended ways that may reduce productivity, create potential information security concerns, or lead to other problems.

Given the growing use and interest in these communication technologies and their potential uses and misuses in the workplace, it is reasonable to expect the development of various guidelines for their appropriate use. Although the growth in information and communication technologies (ICTs) has led to increased development of policies governing communication, it is not clear how extensive or adequate those policies addressing mobile communication are.

This chapter first seeks to establish the importance of workplace policies related to mobile communication. We then describe the methods used to analyze relevant technology policies from U.S. state governments—which allows us to meet our primary objective of describing mobile communication policies in terms of breadth and depth of various themes/categories. We use the findings to offer scholarly and practical implications.

Technology, Policy, and Organizations

Townsend and Bennett (2003) note that policy matters because it protects the organization from legal problems associated with system misuse and abuse, encourages appropriate use, enhances productivity, and discourages illegalities. A similar set of policy benefits for employees and other organizational members could also be listed. Most of the interest in what is broadly called communication and information policy research has centered on "how regulators, governments, and public policies shape communication-information industries and social practices" (Mueller & Lentz, 2004, p. 155). However, large-scale policy research in that sense has not examined policies in the more local for-profit, nonprofit, or government workplaces where communication policies are developed and enforced. Work-related policies represent a form of organizational communication and a number of these policies specifically address topics related to ICTs—commonly taking the form of privacy policies (see D'Urso, 2006; Townsend & Bennett, 2003), telecommunication polices (see Whitman, Townsend, & Alberts, 1999), and especially acceptable use policies that explain what the organization sees as appropriate technology use (see Barnes, 2002; Gaskin, 1998; Simbulan, 2004). Concerns about issues such as cyberslacking and records retention have resulted in various ICT policies becoming increasingly common in the workplace (American Management Association, 2005).

Organizations have now begun to actively draft such policies, partly in response to perceived misuse of such tools in organizations (Zetter, 2006). Indeed, Simmers (2002) suggests explicit and clearly communicated policy is key in addressing issues related to loss of intellectual property, productivity losses due to online activity, security threats, and network overload. The 2005 Electronic Monitoring and Surveillance Survey from the American Management Association (2005) revealed that 84 percent of organizations have written policy governing personal email use; another 81 percent have written policies about personal Internet use.

Thus, organizational policies specifically related to ICT use have become increasingly common in recent years—especially for technologies such as email and Web browsing. Although research on workplace policies from any organizational perspective remains relatively limited, it does exist (see Pollach, 2006; Siau, Nah, & Teng, 2002; Simbulan, 2004; Simmers, 2002; Ugrin & Pearson, 2008). Far less frequent is anything about mobile phones and other portable devices, to which we turn next.

Mobile Communication Policy

The previously mentioned 2005 survey found that 27 percent of organizations have written policies concerning personal cell phone use at work; 19 percent use policy to control the capture of images from camera phones (American Management Association, 2005). That survey reported 6 percent of organizations have fired employees for misusing phones and another 22 percent have issued major reprimands to individuals abusing phone privileges. The association's 2007 survey finds 45 percent of organizations monitor time spent and numbers called; another 16 percent and 9 percent, respectively, record phone conversations and monitor voicemail (American Management Association, 2008).

The rationale for such policies centers in part on concerns about appropriate use of such technologies. Ugrin, Pearson, and Odom (2007) reported that employee abuse of the Internet in the workplace has become a pervasive problem. Such misappropriation of organizational resources can put the organization at risk (Nolan, 2005), and one of the best ways to avoid these pitfalls is to develop and implement an effective acceptable usage policy (AUP). However, despite the proliferation of mobile devices and the development of those policies, it is not clear how extensive or adequate those policies are.

Indeed, few scholarly studies have examined policies related to mobile communication technology use. Some of these are focused on broader social policies related to mobile phone use, such as licensing issues (see Schejter, 2006). Others have looked primarily at policy challenges linked to the use of mobile devices in various public spaces (Wei & Leung, 1999); although policy was offered as a solution for problematic use, workplace policies were not a focus.

Two other studies make more explicit links between workplace policies and mobile communication technologies. Barker, Cobb, and Karcher (2009) discuss record retention policies with specific mention of cell phones and PDAs as devices where such records are increasingly held. As they note, "Even those organizations that have robust policies probably have not yet begun to deal with a wireless, virtual workplace" (p. 183). Lever and Katz (2007) looked at existing policies related to mobile phone use in libraries based on a national sample of colleges/universities. They found cell phone control policies in the form of online/Web listings, in signage, in online/signage combinations, in operating procedures/manuals, and some with no policy at all. Policies also ranged in terms of topics covered (noise level, where allowed to use) and enforcement levels.

Research Questions

Mobile communication policies are almost certainly needed in the workplace. Though not as widely developed as general Internet and email policies, they do appear to exist; however, we know little about this from scholarly research. This, along with the growing development of ICT policies in the workplace generally, suggests it is time to more closely examine mobile communication policies in organizations.

RQ1: To what extent do policies addressing mobile communication issues exist in the workplace?

RQ2: What specific themes and broader categories are found in workplace mobile communication policies?

Method

Data Collection

The policies analyzed in this research were publicly available on the official websites for each of the U.S. state governments (including the District of Columbia). As Sharf (1999) noted, the Internet is a potentially excellent source of data for researchers wishing to examine discourse itself. The data collection for this study was part of a larger policy-based research project in spring 2008 conducted at a major research university where publicly available ICT policies on a number of research topics were collected for later analysis. State governments seem an appropriate context for such work because (a) policy may be better developed here given constitutional protections applicable to public employees, (b) they represent sizable employers who have historically made use of ICTs, and (c) as a public entity their policies are more readily available online.

Our identification of the relevant policies focused on statewide policies applicable to all state employees using ICTs. The broader project captured policies and other guidelines/resources about terms including Internet, email, IT, technology, information, online, phone, wireless, telecommunication, and communication. That work produced nearly 300 different policies (approximately 6 per state).

Analysis

The policy data was further analyzed in two steps. First, coauthors divided the states and searched for the following words, word stems, and phrases—excluding references that were clearly irrelevant: mobile, wireless, portable, cell, phone, laptop, pager, PDA, personal digital assistant, BlackBerry, provider, and personal use. We also searched for mobile device functions: call, text, game/gaming, messaging, camera, picture, music, and calendar. The relevant excerpts—usually a numbered portion of a policy document or a paragraph or broader section—were then extracted and identified by state and policy document name. This automated coding was done to help focus attention on relevant sections of a vast number of policies.

Second, the authors analyzed the relevant policy excerpts using a constant comparative method consistent with grounded theory development (Glaser & Strauss, 1967). After initially deciding to take a broad view of policy themes that might include both general and specific topics addressed, we then divided the policies and began examining them separately. Focusing on the language around the highlighted terms from step 1, each researcher began a system of open coding to identify salient themes that related to some aspect of mobile communication. The researchers individually produced several overlapping and multiple non-overlapping themes. Joint discussion of the themes, which also involved a return to the data to compare themes to policy statements, helped produce the final set of policy themes. In one final step, the authors considered the individual themes for broader *categories* of policy themes.

Findings

Research Question 1

We found 36 state governments having policy pertaining to mobile communication. The first step of coding produced 146 single-spaced pages of relevant policy excerpts (averaging 4 pages/state, but ranging from 1-23 pages). On average, each state included 2.4 (range 1-14) different policies with relevant information. Even when there was similar coverage of issues, some states would group all that under one or a few policies whereas other states provided multiple different policies. Table 10.1 provides names of the various policies we found addressing issues related to mobile communication.

Table 10.1 Names of State Policies Addressing Mobile Communication Topics

Acceptable Use	Malicious Code Security
Acceptable Use of Electronic	Mobile Data Standards
Communication Devices	Mobile Device
Acceptable Use of Portable Devices	Mobile Storage Devices
Acceptable Use of State Provided	Network and Telecommunications
Wireless Devices	Network Security Standard

(*continued on next page*)

Table 10.1 (*continued*)

Access Control	Personal Computer
Accessibility	Portable Computing Security
Cellphone Use	Remote Access Security
Cellular Devices and Services	Required Notification of Actions
Collaboration Technology Standards	Involving Wireless Communication
Data Classification	Infrastructure
Email Acceptable Use	Security Education and Awareness
Emergency Telework	Security Incident Response
Enterprise Security Policy on Portable	Software Licensing
Computing Devices	State Agency Uses of Cellular Telephones
Incident Management	System Access Control
Information Security	Telecommunications Equipment
Information Technology Standard	Telecommunications Utility Services
Internet Access	Telework
Internet Security	Use of Internet, Email and Other
Intrusion Prevention and Detection	IT Resources
IT	Use of State Telephones
IT Resource Acceptable Use	Web Accessibility
IT Security Incident Reporting	Wireless Cellular Data Technology
Laptop and Remote Access Acceptable	Wireless Communication
Use Guidelines	Wireless LAN
Laptop Encryption	Wireless Telephone Equipment and
Loss Control Policies, Procedures and	Service Usage
Practices	Wireless Access Point Software

Table 10.2 provides frequency counts for each of the key policy terms we examined. Clearly, the most frequently mentioned words in these policies are wireless, phone, call, cellular, and mobile. Less frequently mentioned are terms related to PDAs, laptops, messaging, and games.

Table 10.2 Frequency of Terms Related to Mobile Communication in State Government Policies

Search Term	Number of States with Document Containing Term	Number of Times Term Used
Cell (Cellular phone)	18	281
Mobile	23	252
Mobile Phone	6	15
Tele(phone)	25	358

Table 10.2 (*continued*)

Search Term	Number of States with Document Containing Term	Number of Times Term Used
Wireless	23	485
Laptop	19	75
PDA(s)	15	39
Game	8	19
Messaging (Text message, SMS, MMS)	15	52
Picture(s)	2	2
Music	4	6
Call(s)	26	292
Personal call(s)	7	32
Phone call(s)	9	22
Personal use	15	67

Research Question 2

Our thematic analysis resulted in 15 themes within and across four broader categories (see Table 10.3). Before describing each of those, we note some general confusion in policies concerning their applicability to mobile devices. In one sense, this applies to broad policies about email and more general Internet use that seem to assume predominantly desktop computers (but are increasingly possible on mobile devices). Indeed some policies note scope limitations such as the following: "the scope does not include State phone systems, fax machines, copiers, state issued cell phones or pagers unless those services are delivered over the State's IP network" [Tennessee]. More specifically, there is clear variation across policies concerning how broadly the term phone applies. A number of policies distinguish "phone" and "telephone" from "mobile phone" and "cellular phone." In these instances, there are separate policies covering different types of devices. However, sometimes a policy will refer to just a phone when clearly describing a mobile device: "drivers of LP vans shall not use a phone while the vehicle is in motion" [North Dakota]. Other states are very explicit: "unless otherwise stated, telephone service includes both wired telephone and wireless telephones" [Ohio]. Thus, we have attempted to infer from context when not otherwise clear if a policy applied to mobile devices. Based on that, we report findings around four broad categories emerging from our themes.

Table 10.3 Categories and Themes Related to
Mobile Communication Policies

Theme Category	Policy Theme
Supplemental/ Background	Scope/rationale
	Definitions
	Reference to other policies
	Technical specifications*
Tracking/Monitoring	Inventories and audits
	Accessing and protecting device
	Monitoring/surveillance
Eligibility	Allocation of mobile devices
	Acceptable wireless providers/vendors
	Technical specifications*
Users/Use	Personal use
	Work use
	Security issues
	Training to use
	Use of the devices in emergency situation
	Notification/Contact Lists

* One theme fit across two categories.

Supplemental/Background Information

One of the potentially less central categories contained several themes that provided background or clarifying material to the policies. In this category are three themes related to scope/rationale, definitions, and reference to other policies—as well as parts of a fourth theme addressing technical specifications.

Several formal policies had specific purpose or scope sections at the outset of the document. These statements might describe either the relevant technologies covered or the employees and others to whom the policy applied:

1.0 Purpose

…The purpose of this policy is to establish and communicate acceptable use, security, and confidentiality requirements related to the use of state-owned portable devices within all sites connected to and supported by the State's network infrastructure….

2.0 Scope

> This policy applies to all State employees who use State-owned and authorized portable devices [West Virginia].

In other instances the scope/purpose was embedded into other sections and a rationale for either the policy that followed or for mobiles more generally:

> Advances in wireless technology and pervasive devices create opportunities for new and innovative business solutions. However security risks, if not addressed correctly, could expose SE information systems to a loss of service or compromise of sensitive information... [South Carolina].

In some states even abstracts, objectives, and/or appendices were provided. All these provide a type of supplement to set the stage and provide rationale for the main policy.

A second theme here concerns definitions. Sometimes these were embedded into policies and terms were defined as they emerged. In other instances, special sections were provided for defining both the technologies involved and even some specific terms (e.g., "essential personal call"):

> The term "mobile computing devices" refers to portable or mobile computing and telecommunications devices that can execute programs. This definition includes, but is not limited to, notebooks, palmtops, PDAs, IPods, BlackBerry devices, and cell phones with Internet browsing capability [Delaware].

A third theme was reference to other policies. These were likely helpful to readers in directing them elsewhere in that policy or to entirely different policies to access needed information:

> User responsibilities specified in policy ENT-SEC-081 User Responsibility and ENT-SEC-041 Transmission Privacy apply in so far as a cellular device under this policy provides a capability listed in those policies [Montana].

In some instances other policies are cited as being consistent with a specific policy, perhaps using the former to grant legitimacy to the latter:

> Consistent with NYS Technology Policies 96-2 and 96-2A... any proposed action involving wireless communications initiatives, as defined in Part A below.... shall be centrally coordinated through the New York State Office for Technology in compliance with this policy [New York].

Finally, our one theme that cut across two categories addressed technical specifications. In some cases technical specifications were used as supplemental and general background information:

> This standard provides guidelines for the hardware procurement and secure deployment of wireless access points for "Hot Spot" Wi-Fi coverage and Point-to-Point LAN communications links covering IEEE specifications 802.11a, 802.11g and 802.11i. ...Many security concerns developed since the original specification of IEEE 802.11, 802.11b and the WEP or RC4 encryption standards [New Mexico].

Tracking/monitoring

Part of what the policies cover fundamentally addresses issues of tracking the technologies. This can be done by the user, but more typically by the organization. One clear theme here was a policy to inventory/audit devices so as to keep accurate records of the tools and their locations as well as their status and readiness for necessary updating:

> Maintain an inventory of all wireless devices that lists each individual device, the service provider for such device and the individual... to which the device is assigned [Georgia].

Some even detail the process of return or disposal of a device and removing it from inventory:

> Agencies shall develop policies and procedures for the return of state-owned portable computing devices when the user's employment or contract terminates, or when the user's assignment no longer requires the state-owned device.... When a device is removed from service, state agencies shall sanitize the IT equipment [Ohio].

A second theme addressed keeping track of the physical device to ensure no unauthorized access by others—either by taking the device or by somehow gaining access to it:

> Mobile computing facilities should also be physically protected against theft especially when left, for example, in cars and other forms of transport, hotel rooms ... should not be left unattended and, where possible, should be physically locked away... [Oklahoma].

A third theme in this category concerns monitoring/surveillance as yet another way we track these technologies and their use. Organizations seem to vary in how explicit they are about such policies. In one extreme example of a right to monitor view, we found the following:

> The State has the right to monitor and review these devices for operational or management purposes.... Employees should have no expectation of privacy regarding their use of wireless devices. The State reserves the right to review calls, voicemail messages, text messages, etc., on all State-provided wireless devices... at any time without the consent, presence, or knowledge of the affected employee. This may include usage, voicemail, text messages, and/or e-mail [West Virginia].

Eligibility

This category contained two unique themes (allocation of mobile devices and wireless providers/procurement) and one theme (technical specifications) shared with another category. Allocation of mobile devices encompasses the eligibility of persons and/or jobs that are suitable to get mobile devices, as well as organizational decisions to use portable devices. Some states even specifically suggest guidelines or criteria state agencies can follow when allocating devices:

> Agencies must establish and adopt written criteria to determine which job functions have a legitimate business need for wireless or mobile devices.... Where communications needs cannot be met with fixed telephone or desktop computer equipment, an agency must use their Rules, Regulations and Procedures Governing the Acquisition and Use of Telecommunications Services and Equipment... [Georgia]

Employees should meet one or more of the following guidelines in order for a wireless communication device to be assigned.

> The duties of the position are such that immediate emergency response is critical to successfully carrying out the job (e.g., police officer, fire or emergency responder, etc.)... [Pennsylvania].

This theme also includes the purpose of assigning mobile devices, and organizational approval requirements when allocating state-owned mobile devices:

> State employees are provided the use of portable devices on an as-needed basis to access the State network, or to conduct State business from a remote location, except as specified within this document [West Virginia].

> State agency division administrators or their designees shall review and approve requests for cell equipment and services consistent with these requirements [Montana].

Another theme under Eligibility is providers/procurement. This refers to the information stating that employees should attain mobile devices through certain designated service providers and follow certain guidelines when deciding who is eligible to provide mobile services:

> Agency wireless telephone equipment and service shall be procured through the mandatory state contract wherever equipment and service is available and through the terms and conditions of that contract... [Missouri].

> Procurement by agencies of wireless communication devices and subscription services is to be in strict adherence with the contractual agreement between the Commonwealth and its wireless technology service providers. No wireless technology shall be purchased from vendors outside the approved statewide contract [Pennsylvania].

Finally, one theme fitting partly in this category concerns technical specifications. Some policies offered specific technical guidelines about wireless standards or other minimum requirements for vendors or purchase of technologies:

> Dell Latitude and Precision Models listed on the State Dell Premier Page (not Inspiron)... 64-bit Windows operating systems are not allowed—802.11 a/b/g support required if install wireless [Alaska].

User/Use Issues

This category consists of six different themes. These include personal use, work use, security issues, training, emergency use, and notification/contacts list. In several ways, these issues get at the heart of the policy content in addressing concerns about the user and the usage of mobile devices.

Personal use encompasses any use of mobile device (especially if state-owned) for purposes other than work, including (but not limited to) personal phone calls and Internet usage. Unacceptable uses of mobile devices include the following:

> Accessing web sites offering online gambling, games, and related information such as cheats, codes, demos, online contests... and sites that promote game manufacturers [Alabama].

There were also some policies related to the use of state-owned phones for personal use—often prohibiting all but "essential" calls and requiring reimbursement to the state:

> The use of state-owned cellular telephone equipment and service is intended for state business. Personal use of state-owned cellular phones is prohibited, except for essential personal calls [Minnesota].

> [D]oes not require state agencies to adopt policies absolutely prohibiting any personal use of telephones or computer services as long as the state is reimbursed for any direct costs incurred [Texas].

Although efforts were often made to define terms like "essential" and "emergency" use, policies were often somewhat vague on these issues, noting only that use beyond "minimal," "excessive" use, and anything that interfered with work was not allowed.

A second theme in this category, work issues, refers to restrictions on use of the technology even for work-related purposes. The policy here ranges widely from concerns about the specific times/places that a device can be used to issues about whether mobile or more fixed technologies are appropriate:

> Employees are strongly encouraged not to use handheld cell phones or other handheld electronic communications devices or objects while operating state vehicles or personal vehicles on state business [Montana].

> Cellular phones are not to be used when a less expensive alternative is available [Mississippi].

Security issues related to use are a third theme. Since these devices are generally state-owned and may be used to transmit sensitive material, we considered confidentiality and privacy practices as themes related to security:

> Employees should ... avoid use of any wireless or cellular phones when discussing
> sensitive or confidential information; not send sensitive or confidential documents
> via wireless fax devices; not send teleconference call-in numbers and passcodes to
> a pager, if sensitive or confidential information will be discussed during the confer-
> ence [South Carolina].

A fourth theme in this category addresses user training. For the most part, this concerned the need for greater education related to various usage practices:

> Training must be provided to staff using mobile computing resources to raise their
> awareness on the additional risks resulting from this way of working and the controls
> that will be implemented [New York].

> Agencies shall ensure that restrictions and controls on personal use of IT resources
> are addressed by education and awareness programs [Ohio].

A fifth theme, use of the devices in emergency situations, was men-tioned only rarely. This includes any mention of the use of devices in crisis scenarios, as well as specific rules for "turning off" the devices in certain emergency situations:

> Make provisions for an emergency communications system such as a cellular phone,
> a portable radio unit, or other means so that contact with local law enforcement, the
> fire department and others can be swift [North Dakota].

As a final theme, notification/contacts lists refer to information on contacts for technical support, emergencies, use, or training:

> All contact information must be available for IT management, team members, all
> IT personnel and designated business unit management. When available, this in-
> formation should include: work telephone number, pager number, home telephone
> number, cellular telephone number, work email address, home email address, and
> home address [Oklahoma].

In some instances, policy dictated not including such information for contact because of greater cost:

> Due to associated costs for outgoing and incoming communications, addresses and
> telephone numbers for wireless devices shall be unlisted and shared for business-
> related purposes only. Employees should not place their cellular telephone numbers
> on their business cards [Pennsylvania].

Conclusions

This study examined the breadth and depth of policies related to mo-bile communication in U.S. state governments. Our analysis suggests substantial variation in policy coverage and notable variation across states. We close the chapter by briefly discussing several key conclu-sions, implications, and directions for future research.

Our analysis points to the existence of substantial relevant policy in these states. However, those standards are spread across a diverse range of policies in fairly uneven ways from state to state. Most notably, rules pertaining to mobile communication can be found in numerous different policies and are rarely consolidated.

A second conclusion is that there is clearly some confusion and ambiguity present in these policies. Some of this is likely helpful in providing room for interpretation and developing reasonable standards—which can be especially valuable in determining what "minimal" or "excessive" personal use entails, for example. In other ways, this ambiguity can be harmful when it is unclear which policies do and do not apply to mobile communication specifically—especially when terms like "mobile" can refer to such a range of technologies. There are additional challenges created in the inconsistency and coverage variation in some policies from state to state—especially when states interact with one another.

Third, we observed a sizable range of issues addressed in these policies. Some were about routine work issues and others about emergency situations. The relevant audiences/stakeholders varied from information technology specialists to purchasers, to users and their supervisors. Again, this variation in topics and audiences creates some challenges for effectively communicating this policy information.

Additionally, we see the use/usage issues as the most central category here given the number of themes it encompasses and the amount of relevant policy covered. Although "personal use" issues are prominent, they are not dominant amid several other user/usage topics. It is refreshing to see so many policies directed here, even if they tend to be more about restrictions than rights.

Finally, we conclude that the supplemental category and themes under it are important even if less central than other categories and themes. These provide both a narrative rationale for the importance of certain technologies/policies and clarification of key terms or other relevant policies. We believe such information is generally useful to those who must utilize policy.

Implications

One of the scholarly implications of this research is that detailed analysis of organizational policy documents can produce important research findings. We see this work as helping legitimize policy messages as a valuable form of organizational communication. Our efforts here also provide a reasonable methodological approach for policy analysis in the workplace. Additionally, this sort of analysis can inform research and theory building related to mobile communication and new media by helping provide better understanding of key rules and resources that may guide mobile technology use. Structuration approaches (Giddens, 1984) would seem quite appropriate here as a way to better examine how policy ultimately shapes and is shaped by use in practice.

There are also practical implications that emerge from this work. Some of these are for policy makers who may see the range of related issues addressed in these policies and the variations that exist across them. Such information can lead to the development of smarter policy. Users may also benefit from this work by highlighting the importance of the available policy information—and thus the need to be aware of the rights and restrictions provided. Supervisors and organizational management can use the data here to reconsider enforcement issues and make organizational members better aware of relevant policies. Our findings do not necessarily suggest more policy is needed, though existing policy could always be improved to address the issues highlighted here.

Future Research

Future research should explore the impact of such formal policies and the degree to which policies are followed by mobile communication users and their employers. Additionally, we have only examined policy in one type of organization here; thus, future research should extend this analysis by examining policy in the private sector and in organizations whose members are more globally dispersed. Future work may also wish to explain differences across states based on geographic, socio-political, or other factors. Comparing workplace policies to broader social policies about mobile communication could help reveal not only important distinctions between the two, but also the potential influences each has on the other.

In closing, mobile communication policies are an increasingly important part of workplace communication—both as a form of communication and as regulator of such interaction. The description of current state policies in this area and the categories/themes contained within have suggested several conclusions and implications for better understanding and theorizing about mobile communication and workplace interaction.

References

American Management Association and the ePolicy Institute, (2005). *Electronic monitoring & surveillance survey.* Retrieved on June 18, 2009 from http://www. epolicyinstitute.com/survey2005summary.pdf.

American Management Association and the ePolicy Institute, (2008, 28 February). 2007 *electronic monitoring & surveillance survey.* Retrieved on June 18, 2009 from http://press. amanet.org/press-releases/177/2007-electronic-monitoring-surveillance-survey/.

Barker, R. M., Cobb, A. T., & Karcher, J. (2009). The legal implications of electronic document retention: Changing the rules. *Business Horizons, 52,* 177-186.

Barnes, S. B. (2002). *Computer-mediated communication: Human-to-human communication across the Internet.* Boston: Allyn & Bacon.

D'Urso, S. C. (2006). Toward a structural-perceptual model of electronic monitoring and surveillance in organizations. *Communication Theory, 16,* 281-303.

Gaskin, J. E. (1998). Internet acceptable usage policies. *Information Systems Management, 15,* 20-25.

Giddens, A. (1984). *The constitution of society.* Berkeley: The University of California Press.

Glaser, B., & Strauss, A. (1967). *The discovery of grounded theory: Strategies for qualitative research.* Chicago: Aldine.

Kelly, T., & Biggs, P. (2007). Mobile phones as the missing link in bridging the digital divide in Africa. *African Technology Development Forum Journal, 4,* 11-13.

Lever, K. M., & Katz, James E. (2007). Cell phones in campus libraries: An analysis of policy responses to an invasive mobile technology. *Information Processing and Management, 43,* 1133-1139.

Mueller, M., & Lentz, B. (2004). Revitalizing communication and information policy research. *The Information Society, 20,* 155-157.

Nolan, J. (2005). Best practices for establishing an effective workplace policy for acceptable computer usage. *Information Systems Control Journal, 6.* Retrieved on June 20, 2009 from http://www.isaca.org/Template.cfm?Section=Archives&CONT ENTID=27801&TEMPLATE=/ContentManagement/ContentDisplay.cfm.

Pollach, I. (2006). Privacy statements as a means of uncertainty reduction in WWW interactions. *Journal of Organizational & End User Computing, 18,* 23-49.

Schejter, A. M. (2006). Israeli cellular telecommunications policy. *Telecommunications Policy, 30,* 14-28.

Sharf, B. F. (1999). Beyond netiquette: the ethics of doing naturalistic discourse research on the Internet. In S. Jones (Ed.), *Doing Internet research: Critical issues and methods for examining the net* (pp. 243-256). Thousand Oaks, CA: Sage.

Siau, K., Nah, F. F., & Teng, L. (2002). Acceptable Internet use policy. *Communications of the ACM, 45,* 75-79.

Simbulan, M. S. R. (2004). Internet access practices and employee attitudes toward Internet usage policy implementation in selected Philippines financial institutions. *Gadjah Mada International Journal of Business, 6,* 193-224.

Simmers, C. A. (2002). Aligning Internet usage with business priorities. *Communications of the ACM, 45,* 71-74.

Townsend, A. M., & Bennett, J. T. (2003). Privacy, technology, and conflict: Emerging issues and action in workplace privacy. *Journal of Labor Research, 24,* 195-205.

Ugrin, J. C., & Pearson, J. M. (2008). Exploring Internet abuse in the workplace: How can we maximize deterrence efforts? *Review of Business, 28,* 29-40.

Ugrin, J. C., Pearson, J. M., & Odom, M. D. (2007). Profiling cyber-slackers in the workplace: Demographic, cultural, and workplace factors. *Journal of Internet Commerce, 6,* 75-89.

Wei, R., & Leung, L. (1999). Blurring public and private behaviors in public space: Policy challenges in the use and improper use of the cell phone. *Telematics and Informatics, 16,* 11-26.

Whitman, M. E., Townsend, A. M., & Alberts, R. J. (1999). Considerations for an effective telecommunications-use policy. *Communication of the ACM, 42*(6), 101-108.

Wray, R. (2008). Half world's population will have mobile phone by end of year. Retrieved on June 18, 2009 from http://www.guardian.co.uk/technology/2008/sep/26/mobilephones.unitednations.

Zetter, K. (2006, October). Employers crack down on personal net use. *PC World,* p. 26.

11

ICT Use and Female Migrant Workers in Singapore

Minu Thomas and Sun Sun Lim

As Singaporean women enter the workforce in larger numbers, thus contributing to a rising proportion of dual-career families, demand has grown for live-in maids who can ease working women's household burdens (Yeoh, Huang & Rahman, 2005). According to recent estimates, at least 150,000 female migrant workers from countries such as Indonesia, Philippines and India are employed as live-in maids in Singapore (Rahman, Yeo & Huang, 2005), constituting over one-fifth of Singapore's migrant workforce. These maids perform household chores such as cooking, cleaning, grocery shopping, child-minding and caring for elderly family members. Unlike migrant women who work in factories or offices and have their own accommodation, maids have relatively little autonomy as they live with their employers. Hence, their access to basic necessities such as food, shelter, communication, medical treatment and the right to private space and time are determined by their employers. Previous research has found that within the employers' homes, maids are treated neither as family members nor as total outsiders (Rahman et al., 2005). In some homes, employers use security cameras and other ICT tools to conduct surveillance on their maids (Au Yong, 2005) and seek to prevent their maids from developing strong ties with their families back home (Yeoh et al., 2005). Outside their homes, employers restrict their maids' movements because they are held liable if their maids go missing or become pregnant (Ministry of Manpower, 2009). Hence, employers limit, as far as they can, their maids' social activities and forays into public space as they fear that their maids may foster romantic or sexual relationships or even engage in prostitution. Employers have been known to withhold their maids' passports, impose curfews and limit their access to communications (Sun, 2006). Apart from fetching their young charges to or from school, running errands or accompanying their employers' families on their outings, most maids spend the bulk of their work days in their employers' homes. Thus, they have few opportunities for an active social life and tend to lead isolated existences in their employers' homes and experience loneliness and depression (Rahman et al., 2005). Given such constrained living and working conditions, these maids' communication with the outside world assumes great importance. Their sense of well-being hinges on their ability to reach beyond the confines of their domiciles to engage in social interaction, seek companionship, solicit help and maintain ties with their loved ones back home. In this regard, information and communication technologies (ICTs) may offer these maids a crucial lifeline as devices such as mobile phones and the Internet enable long-distance communication with their friends and kin.

Singapore is a highly networked society where ICT use is intensive, whether in governance, business, education, public service or interpersonal communication. As of April 2008, household broadband penetration was 82.5 percent and mobile phone subscriptions had risen to over 5.9 million, exceeding the national population of 4.6 million (Infocomm Development Authority [IDA], Jan-Jun 2008). Yet such ICT access may not be readily forthcoming for low-waged migrant workers such as maids, who may lack the autonomy, resources or skills to avail of ICTs in their personal communications. If they can and do use ICTs, which do they use most avidly, what benefits do they derive from its use and what challenges do they encounter? To answer these questions, our study sought to understand whether and how maids in Singapore use ICTs in their daily lives and the impact of such use.

Related Literature

A growing body of extant research has been conducted on migrant workers' ICT use and its implications for connectivity and empowerment. Vertovec (2004) argues that cheap international phone calls, despite their prosaic nature, serve as a crucial social adhesive which binds communities around the globe, especially non-elite transnational migrants. Indeed, for many migrant workers, phone calls are indispensable for providing social support to and staying in touch with family members who are working overseas (Strom, 2002). Beyond phone calls, migrant workers have varying levels of access to mediated communications including ethnic or local mass media which serve immigrant communities, transnational mass media such as mobile phones which facilitate mediated interactions across transnational time–space contexts and global media such as the Internet which enable quasi-mediated interactions for global audiences (Benitez, 2006). Studies of migrant workers in different parts of the world indicate that they employ a wide range of ICT devices and services in their communications, whether local or transnational, with different ICTs dominant for particular regions or communities. For example, El Salvadoran migrants in the US use mobile phones, prepaid cards, videoconferencing and home videos to keep in touch with their significant others (Benitez, 2006). In Jamaica, transnational families rely on the mobile phone to keep in touch with overseas friends and kin (Horst, 2006; Horst & Miller, 2006.) In South China, working class migrants depend upon online chats at cybercafés, payphones, Little Smart (a geographically confined but inexpensive wireless service for communication within cities) or prepaid mobile services (Chu & Yang, 2006; Lam & Peng, 2006; Qiu, 2007, 2008a, 2008b). This ability to avail of different communication services can be gratifying and empowering for migrant workers. Chinese migrants find the mobile phone indispensable for job opportunities, maintaining contact with kith and kin, expanding their social networks and engaging in courtship (Lam & Peng, 2006; Strom 2002; Qiu, 2007, 2008a, 2008b). Amongst Jamaican migrants, the mobile phone has become particularly expedient for arranging for remittances and obtaining money during emergencies (Horst, 2006).

Conversely though, ICT use can also be burdensome for migrant workers. The technological divide between the home and host countries of migrant workers means that communication is not always seamless and problem-free (Benitez, 2006). At the same time, the skills required to use different communication devices within divergent technological systems across geographical boundaries, to cope with the breakdown of devices and faculty services and to select the most economical pricing plan for one's communication needs can be especially challenging for transnational migrants from developing countries (Panagakos and Horst, 2006). For example, Qiu (2008) observed that the design of inexpensive mobile phone handsets targeted at Chinese working class migrants omitted basic functions that consequently made them difficult to use. He also found that while young migrant workers in South China enjoyed a high level of ICT connectivity, they were also the most likely to lose control over their ICT budgets. Similarly, Law and Peng (2006) noted that Chinese migrant workers in Guangdong were so accustomed to the perpetual contact that the mobile phone enables that they spent a disproportionately large part of their salaries on mobile phone services, and even more on social activities arising from mobile communications.

A subset of the research on migrant workers' ICT use has focused on transnational families, of which some delve specifically into the experiences of transnational mothers. In a study of migrants and refugees in seven countries, it was found that different communication modes had different implications for how transnational family relationships were sustained (Wilding, 2006). Email helped to improve the quality and quantity of communication while cheap and instantaneous communication via phone, fax, and email enabled them to participate in their family members' lives. The study also found that while ICTs help to mediate a sense of togetherness, it paradoxically intensified the feeling of distance because the intimacy of long-distance contact made the lack of face-to-face contact even more palpable. As for migrant mothers in particular, Uy-Tioco (2007) found that the mobile phone enabled female Filipino migrant workers in the US to remotely assert their roles as mothers. They reinforced their love for their children through text messaging and maintained their presence at home despite the geographical distance. In the same vein, Parrenas (2005) observed that in contrast to migrant fathers who were inclined to maintain only instrumental communication with their families, migrant mothers sought to foster intimate ties with their children through regular communication via phone calls, text messages or letters. Besides the fostering of emotional bonds, transnational families have been found to use ICTs for instrumental communication including arranging for or advising on remittances and providing guidance on their children's homework (see for example Horst and Miller, 2006; Parrenas, 2005 and Pertierra, 2006).

In light of the groundwork laid by extant research on migrant workers' ICT use, our study sought to address the following research questions:

1.   How do maids in Singapore use ICTs in their everyday lives?
2.   What consequences does ICT use have on their living and working conditions?
3.   What difficulties, if any, do they encounter in their access to and use of ICTs?

Methodology

For this exploratory study, we conducted ethnographic interviews in 2008 with twenty maids working in Singapore—ten each from India and the Philippines. The interviewees were recruited through snowball sampling which is useful for recruiting subjects who are otherwise difficult to locate, such as homeless individuals, migrant workers or undocumented immigrants (Babbie, 2004). The Filipino interviewees were recruited at Lucky Plaza on Orchard Road where Filipino migrant workers, including maids, congregate on Sundays. The interviews with these Filipino respondents were also conducted at these informal gathering spots. The Indian respondents were recruited through the researchers' personal contacts, including employers of maids. Hence, the Indian respondents were either interviewed at their employers' homes or at food courts nearby. Given the circumstances, the Indian respondents may have been less forthcoming during the interviews. Be that as it may, all interviewees were given firm assurances about the confidentiality and anonymity of their views to put them at ease. As the interviewees were uncomfortable about the interviews being audio-recorded, verbatim notes were taken and transcripts were prepared within 24 hours of the interviews being conducted.

For background information, we asked the interviewees about their families, their cities of origin, their age and educational qualifications. We asked them to describe, on a day-to-day basis, which ICTs they used, for what purposes, and whether they derived any gratifications or encountered difficulties from such use. We also enquired as to whom they communicated with on a regular basis, what they communicated about and which communication modes they used, e.g., letter, mobile, landline, computer etc. A profile of the interviewees and the ICTs they used is at Table 11.1, with pseudonyms being used in place of real names.

Table 11.1 Interviewee Profile

Name	Age	Educational status	Years overseas	Technologies used
Indian interviewees				
Jameela	21	Secondary (7th Standard)	1.25	Mobile phone (voice calls only)
Malar	45	Primary (2nd Standard)	15	Landline
Mallika	27	Primary (3rd Standard)	6	Mobile phone (voice calls only) and landline
Mercy	36	Higher Secondary	0.25	Landline
Mita	23	Bachelors Degree	0.5	Mobile phone (voice calls and texting) and letter
Parvati	47	Primary (5th Standard)	15	Landline
Sumathi	32	Secondary (7th Standard)	18	Mobile phone (voice calls and texting)

(*continued on next page*)

Table 11.1 (*continued*)

Name	Age	Educational status	Years overseas	Technologies used
Teena	37	Higher Secondary	2.25	Mobile phone (voice calls and texting) and landline
Tessy	39	High School	10	Mobile phone (voice calls and texting) and computer
Thenmozhi	33	High School	12	Mobile phone (voice calls and texting) and landline
Filipino interviewees				
Abigail	39	High School	11	Mobile phone (voice calls and texting), music player
Candy	22	Bachelors Degree	0.25	Landline and letters
Dale	45	High School	10	Mobile phone (voice calls and texting), music player
Florence	29	Bachelors Degree	2	Mobile phone (voice calls and texting), letters and cards, and computer
Ida	28	Bachelors Degree	8	Mobile phone (voice calls and texting) and landline
Mara	27	High School	2	Mobile phone (voice calls and texting) and letters
Pearl	27	High School	5	Mobile phone (voice calls and texting), letters and computer
Thea	24	Bachelors Degree	1.5	Mobile phone (voice calls and texting), landline, letters and computer
Vera	34	Computer Secretarial Course	15	Mobile phone (voice calls and texting) and landline
Violet	46	High School	19	Mobile phone (voice calls and texting) and computer

Notes: High School indicates 10 years of schooling and higher secondary indicates 12 years of schooling.

Findings and Discussion

While the maids employed a variety of technologies for their communication, including the mobile phone, landline, computer and the Internet, as well as letters and greeting cards, the mobile phone was clearly the most crucial communication device for most of them. Seven of the 10 Indian maids and 9 of the 10 Filipino maids we interviewed owned mobile phones. When they had first started work-ing in Singapore, most of them did not own personal mobile phones but gradually saved up for them. Two of them had been given mobile phones by their employers so that the latter could contact them with instructions or occasionally conduct remote surveillance if the maids left the house on errands or during days off. For these two women, their employers imposed conditions on when and how the phones were to be used, such as restricting the frequency and duration of calls and text messaging, or imposing a prohibition on international calls which were more expensive. In which case, the maids resorted to calling cards to make these international calls.

Ritual and Escape

A typical day in a maid's life begins with household chores such as making breakfast and lunch for her employers, sending the children to school and performing household chores such as cleaning and doing the laundry. By noon they may have some time for themselves, when they can send text messages or call their families and friends. After lunch, their routine resumes with making tea and dinner followed by cleaning, and their day only ends after their employers have gone to bed. As they tend to be exhausted by the end of the day, sleep rather than recreation is their main priority at nighttime.

Ling (2008) opines that the mobile phone is widely used for the per-formance of ritual activities. In the case of maids, their daily routines are highly repetitive and our interviewees eased their drudgery through some private rituals enabled by their mobile phones. Specifically, interviewees who owned personal mobile phones would steal idle moments in their day to send a text message or to call their loved ones, with some maids doing so on a daily basis. The ability to perform these personal rituals of com-municating with their family and friends offered these women a much-needed reprieve from the monotony and tedium of their jobs. Compared to other diversions such as watching television or listening to music, mo-bile phone use was the most convenient and gratifying because it could be used discreetly, and it enabled a precious link to family and friends:

> I use my mobile phone mainly for text messaging. I send around 60 messages a day both to family and friends all over the world. I call my parents and sister twice a week and speak to them for half an hour (Pearl, 27, Filipino).

> I would like to thank the inventor of the mobile phone. I find the mobile most use-ful when I fight with my husband and have to make up with him. I fight with him almost every morning and then I call him from the mobile at least ten times to make up with him (Sumathi, 32, Indian).

It was through these ritual communications that the women in our study could maintain close relationships with family members and friends. Hence, even though their primary role within their employer's home was to serve as the housekeeper, they were able to maintain their personal lives and relationships, and to perform their roles as wives, mothers, daughters, sisters and friends.

Morley (2000) argued that "communication technologies can function as disembedding mechanisms, powerfully enabling individuals (and sometimes whole families or communities) to escape, at least imaginatively, from their geographical locations" (pp. 149–150). While it has been argued that the mobile phone enables people to maintain an absent presence (Gergen, 2002), for the women we studied, the mobile phone was crucial for facilitating a sense of absence from their physical present. Were it not for these personal phone calls or text messages, their communication would mainly be with their employers since they were mostly housebound. With the mobile phone, they could "escape" and maintain an existence, however fleeting and intangible, which extended beyond their lives as maids. In so doing, they were not merely defined by their employment but could regard themselves as individuals with personal goals that transcended the physical confines of their employers' homes. A particularly poignant case was that of 21-year-old Aiysha from India: "*[The mobile phone] gives me a sense of control over my own life and I can at least talk to my family back home in India.*" Her estranged husband had forcibly taken their two children away from her and her communication with her family was crucial in helping her to recover from her personal trials. At the time of the interview, she had just moved into a new employer's home and a sympathetic acquaintance had bought her a mobile phone. When she had first migrated to Singapore to work as a maid, her employer had not allowed her to access any mode of communication: "*I felt cut off from the outside world during those days,*" said Aiysha, who was heartened by her newfound ability to connect with the outside world.

Sociality and Companionship

Katz opined that "[b]y allowing people to transcend a variety of physical and social barriers, the telephone has led to a complex set of dispersed personal and commercial relationships" (p. 116). This was indeed the case for our interviewees and their mobile phone use. Apart from their communication with family and friends back home, the women we studied also sought to grow their social networks beyond and *in spite of* their physical, social and temporal constraints, primarily through their mobile phones. The Filipino workers enjoyed more days off compared to the Indian workers—another factor that aided their more frequent socializing. Six of the ten Filipino workers interviewed were getting a weekly day off, two a fortnightly day off and the remaining two got a monthly day off. As for the Indian workers, three were getting a weekly day off, six got a monthly day off, while one was not getting any days off. The mobile phone played an important role in helping them to coordinate their appointments and other social activities with their friends, mostly through text messaging. Throughout the rest of the week, although they did not get the time to talk or text message on a regular basis, they would keep in intermittent contact with their "local" friends. In this way, the companionship they shared on their days off could be extended, through mobile phone communication, to other days of the week as well:

> When I have to meet up with friends, and I cannot find them at the designated spot, then the mobile phone comes in handy. It also serves as an extension for communication with my friends in Singapore after my weekly day-off (Dale, 45, Filipino).

It was also through the use of ICTs that some workers could take control of their personal lives in terms of initiating romantic relationships and seeking spouses. Despite being women of marriageable age, these women's opportunities to meet eligible life partners were few and far between given the nature of their work. Through ICT-mediated communication however, two of our interviewees were attempting to find life partners. Tessy, a 39-year-old Indian worker had been abandoned by her husband and moved to Singapore to support her children who were back home in India. In due course, she managed to find a life companion by using the Internet and the mobile phone. Her employer had posted her profile on Shaadi.com, an Indian matrimonial website, through which she found a suitable match. In the early days of their courtship, she would use the Internet and webcam at her employer's home to get to know him better. After this initial correspondence, she and her fiancée moved on to text messages and voice calls through the mobile phone to develop their relationship. Mercy, 36, another Indian worker, had had a similar experience. Clearly, for these two women, ICTs such as the Internet and mobile phone "brought the outside world in" and made it possible for them to engage in mediated sociality and to extend their social networks despite being circumscribed by the nature of their employment. These women were therefore able to initiate and maintain relationships in an upfront manner, rather than resorting to furtive dalliances which foreign domestic workers have been documented to conduct on their days off (Yeoh & Huang, 1998). However, it was also with the support of their employers that these two women were able to engage in online dating. Without which, the two women would have been unlikely to have found the time or to have possessed the requisite ICT skills to avail of online dating services.

Contacts and Capital

Beyond emotional needs, the women also had practical and instrumental needs that could be served in part by ICTs, but principally the mobile phone. Horst and Miller (2006) assert that the mobile phone, "far from being peripheral or an additional expense, is actually the new heart of economic survival" (p. 108) because it enables access to one's extended support network in times of need. Indeed, the women we studied had varied needs for which they tapped into their family and friends both at home and in Singapore, as well as their employment agencies and former employers. Our interviewees used their mobile phones to obtain information about, amongst others, remittance methods, discounted airline tickets and job opportunities, thus managing their personal affairs with greater autonomy and efficiency:

> I have been in Singapore for 15 years. I do not write letters or use the landline. For many long years, I have been relying on the mobile phone. My previous jobs were secured through my network of friends connected through the mobile phone (Vera, 34, Filipino).

> My employer gave me some contact numbers of bank managers in India and with this information, I could talk directly to these managers and find out the best ways to invest my money in something like a mutual fund. It is better than just keeping my money locked up in the cupboard or sending it to my relatives in India (Mallika, 27, Indian).

The network of contacts that could be grown and activated through the mobile phone translated into social and economic capital (Bourdieu, 1986) for these women, which in many circumstances, further translated into enhanced capital for their families back home. In transnational movements, "people's social locations affect their access to resources and mobility across transnational spaces, but also their agency as initiators, refiners and transformers of these locations" (Pessar and Mahler, 2003, p. 817). The shifts in the communication modes of our respondents as their overseas stints progressed reflect their improved access to communication resources and their ability to provide enhanced access to communications for their families back home. Thenmozhi's experience illustrates this point. She had first migrated to Singapore in 1996, when there were few public telephones in her village and the cost of international calls was prohibitively high. In the initial months of her stay in Singapore, she used to call her children once every 15 days from public telephones using calling cards. Since her home in India did not have a landline, her children received her calls through a public telephone near their home. At the time of the interview however, Thenmozhi could call her children from her mobile phone daily because she had also bought a mobile phone and a landline for her family to receive her calls. Thenmozhi's income as a domestic worker and her exposure to ICT use in Singapore had given her the technological and financial wherewithal to improve communication between herself and her family:

> I speak to my three children daily for at least five minutes each time. I even keep track of their daily schedules and the mundane occurrences in their everyday lives. I myself received only ten years of education but I want my children to be better educated than me. So, I call them whenever they have exams or any other important event in their lives to encourage them (Thenmozhi, 33, Indian).

This created a virtuous cycle because quite apart from facilitating more frequent contact between mother and children, Thenmozhi was better able to manage her family remotely, thus enabling her to continue with her overseas employment. This in turn put her in an advantageous financial position, bringing the family closer to its goal of attaining higher education for the children.

Burdens and Challenges

While ICTs could serve as instruments of empowerment and connectivity for the women we studied, they were not without burdens and challenges. With greater connectivity, the workers were bound by responsibilities to their family members, especially to children whom they had left behind, and this took a considerable emotional toll on these women. Very often, these women were the first point of contact during a family emergency as they could be relied upon to provide monetary support. Hence the enhanced connectivity, while a boon, also added an emotional burden for these women because they felt obliged to look after their loved ones remotely as they continued to play the roles of mother, wife, daughter and sister.

Another downside to the enhanced connectivity was the challenge of coping with the financial costs involved. The average monthly income of the Indian workers interviewed for this study was S\$321, and the average monthly income of the Filipino workers interviewed was S\$470. As a proportion of their monthly incomes, communication expenses were 12.1 percent for Indian workers and 10.4 percent for Filipino workers (see Table 11.2). These women's restraint and financial prudence was therefore constantly tested as they monitored their communication expenses:

> The mobile phone can be addictive like smoking, especially when I am idle. The result is that I end up spending more money than I expect to (Violet, 46, Filipino).
>
> I use the mobile phone as well as the landline at my employer's home. My mobile bill itself works out to S$50 and added to that I use two calling cards a month to speak to my family thrice a month (Teena, 37, Indian).

**Table 11.2 Average Monthly Income
and Monthly Expenses on Communication**

Income and expenses	Indians	Filipinos
Average monthly income, in Singapore dollars	321	470
Average monthly communication expenses, in Singapore dollars	38.9	49.06
Communication expenses as % of monthly income (average)	12.1	10.4

Policy Implications

This study has sought to contribute to a better understanding of the benefits and costs of ICT use amongst transnational migrant workers. To the best of our knowledge, this is one of the few studies ever conducted on the use of ICTs by live-in maids who have particular living and working conditions as compared to migrant workers who work in factories and offices. Serving as maids in their employers' homes, the lives of these migrant women revolve around those of their employers. Their employers' lifestyles, routines, practices and need dominate and take precedence. Without much autonomy, free time or personal space, it is challenging for these women to maintain or even possess a sense of self-worth and personal identity. The particular circumstances of such employment, which translates into a highly circumscribed and isolated living and working experience for these women, makes the access and use of communication technologies even more crucial. Our findings strongly suggest that ICTs, especially the mobile phone, are indispensable for these women and their sense of well-being. With their mobile phones, the women we interviewed could keep in perpetual and intermittent contact with their family and friends, seek companionship and grow their social networks, thus helping to enhance their social and economic capital.

In most developing and even developed countries, women face several disadvantages owing to social, economic and cultural factors. Amartya Sen who described development as a "process of expanding the real freedoms that people enjoy" (1999, p. 3), pointed to the existence of severe female disadvantages in health and other aspects of well-being in many areas of the Third World. The greater deprivation of females, Sen noted, is linked to relatively low levels of female literacy and absence of social and economic empowerment. In the age of the network society, gender-based inequalities have also come to be characterized by the lack of access to information and knowledge networks. Therefore, ICTs can play a useful role in building information networks and empowering women who are otherwise disadvantaged (Nath, 2001).

Given the irreversible trend of globalization and the growing movement of labor across developed and developing parts of the world (Brown, 2006), it is imperative that we consider the communication needs of transnational workers who leave behind their families in search of better economic opportunities overseas. As more families are "broken up" in this globalised economy, the demand for maintaining ties across the miles will only increase. Mobile communication in particular, has become so commonplace and integral to the everyday lives of people from all social strata that social policy pertaining to migrant workers needs to take into account such realities. Telephone companies have responded to the particular needs of low-waged migrant workers through mobile phone subscriptions or services that are inexpensive and easy to maintain [see, for example, Koh (2009)]. With such options available, migrant workers find mobile communication less prohibitive and are better able to keep in touch with their loved ones and manage their affairs back home.

Social policy needs to keep pace with market innovation. Our findings suggest a few areas for policy intervention. First, while many of the maids we interviewed enjoy enhanced communication as a result of their ICT use, not all of them have the requisite skills or resources to do so. Therefore the training of new maids should, besides equipping them with housekeeping and child-minding skills, also educate them about the access, use and cost of different communication devices and services. Indeed, all migrant workers should be given such training upon their arrival in their host countries, regardless of which industries they join. Second, although the recommended standard contract between employers and maids in Singapore includes a clause stating that "External communications shall be made available" (Association of Employment Agencies Singapore, 2009, p. 16), it should be more specifically worded. "External communications" can refer to asynchronous or less efficient communication methods such as letters. There is also no clear stipulation on the frequency with which such external communications are to be facilitated. Instead, contracts between employers and migrant workers in general should have clear provisions for the employees' rights to communication and specifically, mobile communications. Third, migrant workers enjoy a higher level of technological standards in their host country than in their home country, thus leading to a transnational technological divide which impedes communication between transnational workers and their loved ones (Benitez, 2006). Governments, non-governmental organizations and the private sector should actively seek to narrow this divide through transnational policies that encourage *inter alia*, the provision of easily accessible and affordably-priced synchronous communication services for migrant workers.

References

Association of Employment Agencies Singapore. (2009). *Standard Employment Contract between Foreign Domestic Worker and Employer*. Retrieved August 20, 2009 from http://www.aeas.org.sg/pdf/sec.pdf.

Babbie, Earl. (2004). *The practice of social research* (10th ed.). Wadsworth: Thomson Learning Inc.

Bourdieu, Pierre. (1986). The forms of capital. In J. Richardson (Ed.), *Handbook of Theory and Research for the Sociology of Education* (pp. 241-258). New York: Greenwood Press.

Brown, Janet P. (2006). *Globalisation in 2020*. New York: Nova Science Publishers.

Chu, Wai.-Chi. & Yang, Shanhua. (2006). Mobile phones and new migrant workers in a South China village: An initial analysis of the interplay between the 'social' and the 'technological'. In Pui-Lam. Law, Leopoldina Fortunati & Shanhua Yang (Eds.), *New Technologies in Global Societies* (pp. 221-244). Singapore: World Scientific.

Gergen, Kenneth J. (2002). The challenge of absent presence. In J. E. Katz & M. A. Aakhus (Eds.), *Perpetual Contact: Mobile Communication, Private Talk, Public Performance* (pp. 227-241). Cambridge: Cambridge University Press.

Horst, Heather A. (2006). The blessings and burdens of communication: cell phones in Jamaican transnational social fields. *Global Networks*, 6(2), 143-159.

Horst, Heather A. & Miller, Daniel. (2006). *The Cell Phone: An Anthropology of Communication*. Oxford: Berg.

Infocomm Development Authority (2008, January-June). *Statistics on telecom services for 2008* (Jan-Jun). Retrieved 5 June, 2008 from http://www.ida.gov.sg/Publications/20080212114723.aspx.

Katz, James E. (2006). *Magic in the Air: Mobile Communication and the Transformation of Social Life*. New Brunswick, NJ: Transaction.

Koh, Yi Na. (2009, March 18). 2,000 foreign workers attend Singtel event. *The Online Citizen*. Retrieved August 20, 2009, from http://theonlinecitizen.com/2009/03/2000-foreign-workers-attend-singtel-event/.

Lam, Pui-Lam. & Peng, Yinni. (2006). The use of mobile phones among migrant workers in Southern China. In P.-L. Law, L. Fortunati & S. Yang (Eds.), *New Technologies in Global Societies* (pp. 245-258). Singapore: World Scientific.

Ling, Rich. (2008). *New Tech, New Ties: How Mobile Communication is Reshaping Social Cohesion*. Cambridge, MA: MIT Press.

Ministry of Manpower. (2006). During Employment: Missing Foreign Domestic Worker. Retrieved August 31, 2009 from http://www.mom.gov.sg/publish/momportal/en/communities/work_pass/foreign_domestic_workers/during_employment/missing_foreign_domestic.html.

Morley, David. (2000). *Home territories: Media, mobility, and identity*. London: Routledge.

Nath, Vikas. (2001). Empowerment and governance through information and communication technologies: Women's perspective. *International Information and Library Review, 33*, 317- 339.

Panagakos, Anastasia. N. & Horst, Heather A. (2006). Return to Cyberia: technology and the social worlds of transnational migrants. *Global Networks, 6*(2), 109–124.

Parrenas, Rhacel S. (2005). *Children of Global Migration: Transnational Families and Gendered Woes*. Stanford, CA: Stanford University Press.

Pertierra, Raul. (2006). *Transforming Technologies, Altered Selves*. Manila: De La Salle University Press.

Pessar, Patricia R. & Mahler, Sarah J. (2003). Transnational Migration: Bringing Gender in. *International Migration Review, 37* (3), 812-846.

Qiu, Jack L. (2007). The wireless leash: Mobile messaging as means of control. *International Journal of Communication, 1*, 74-91.

Qiu, Jack L. (2008a). Wireless Working-Class ICTs and the Chinese Informational City. *Journal of Urban Technology, 15* (3), 57-77.

Qiu, Jack L. (2008b). Working-class ICTs, migrants, and empowerment in South China. *Asian Journal of Communication, 18*(4), 333-347.

Rahman, Noor. A., Yeoh, Brenda. S. A., & Huang, Shirlena. (2005). Dignity overdue: Transnational domestic workers in Singapore. In Shirlena Huang, Brenda S. A. Yeoh &, Noor A. Rahman (Eds.), *Asian women as transnational domestic workers* (pp. 231-259). Singapore: Marshall Cavendish.

Sen, Armatya. (1999). *Development as Freedom*. New Delhi: Oxford University Press.

Strom, Georg. (2002). The telephone comes to a Filipino village. In J. E. Katz & M. A. Aakhus (Eds.), *Perpetual Contact: Mobile Communication, Private Talk, Public Performance* (pp. 274-283). Cambridge: Cambridge University Press.

Sun, Shirley. H.-L. (2006, August). *Cellphone Usage and Everyday Resistance of Live-in Maids in Singapore*. Annual Meeting of the American Sociological Association, Montreal, Quebec, Canada.

Uy-Tioco, Cecilia. (2007). Overseas Filipino Workers and Text Messaging: Reinventing Transnational Mothering. *Continuum, 21*(2), 253-265.

Vertovec, Steven. (2004). Cheap calls: the social glue of migrant transnationalism. *Global Networks, 4*(2), 219–24.

Yeoh, Brenda S. A. & Huang, Shirlena. (1998). Negotiating public space: Strategies and styles of migrant female domestic workers in Singapore. *Urban Studies, 35* (3), 589.

Yeoh, Brenda. S. A., Huang, Shirlena & Rahman, Noor A. (2005). Introduction: Asian women as transnational domestic workers In Shirlena. Huang, Brenda. S.A. Yeoh, & Noor A. Rahman (Eds.), *Asian women as transnational domestic workers* (pp. 1-17). Singapore: Marshall Cavendish.

12

Can You Take It with You? Mobility, ICTs and Work-Life Balance

Tracy L. M. Kennedy,
Barry Wellman, and Julie Amoroso

Work Becomes More Mobile and Goes Home

A host of forces in North America have been making work more mobile. Accelerating the trend is the corporate shift in developed countries from making, growing, mining and transporting things—atom-work—to selling, describing, and analyzing things via words and pictures—bit-work—ideas expressed in words, pictures or videos. To be sure, people have worked at home forever: remember Silas Marner weaving at home in the early nineteenth century, shopkeepers living above stores, or farmers living next to their barns. But the dominant organizational paradigm for the past 200 years has been of people commuting from home to work at offices or factories.

The advent of the Internet has made it easy for white-collar workers—bit users—to move around with their work. Depending on their work structure, they can take their laptops on the road to clients and to coffee shops. Our focus is on another form of work mobility—working at home: over-time, part-time and full-time. It is part of the reconfiguration of work from being bound up in closely supervised, physically compact groups to being networked—where people are individually responsible for their own production. The logic of networked work leads directly to mobile work. If people are working in multiple teams, there often is no reason for them to sit side by side. Much work involves writing, drawing, or data analyzing. With the availability of mobile computing and broadband Internet at home, it is possible to work anywhere at anytime.

And what better place for most than their homes—as long as they can get it done without tensions with their domestic lives.

In such situations, the workplace often moves to the home. People finish off tasks that they could not finish at the office, or part of our employment involves working from the home office. Either way, work life often creeps into home life—a quick email or text message to work colleagues between a load of laundry or cutting the lawn.

But what kind of home-based work is actually going on? Is it the full-time "telework" that has fascinated pundits and management gurus since the mid 1990s? There has been more exhortation than research. Some gurus saw telework as the basis of a new form of networked work, in which the constraints of distance would disappear. For example, *Economist* commentator Frances Cairncross heralded *The Death of Distance* in 1997, arguing that:

> New communications technologies are rapidly obliterating distance as a relevant factory in how we conduct our business and personal lives.... The story today is not only the diminishing importance of distance, but also the mobility and ubiquity of technology (back cover).

Companies such as Bell Canada have seen telework as the answer to reduced real estate costs and reduced travel costs (Dimitrova, 2003; Gordon, Gordon & Kelly, 1986; Gordon, 1987). While there has been a substantial movement of work from offices to home, much of it has not been the kind of telework that the gurus foretold. In 2000, approximately 1.4 million people in Canada—just over 10 percent of the population—reported doing some or all of their paid work at home; a 4 percent increase since 1990 and a 1 percent increase from 1995 (Statistics Canada, 2007). Many of today's workers can choose where (and when) they work, primarily because of the affordances that information and communication technologies *(ICTs)* provide—mobility. In 2005, Canadians worked at home occasionally or brought work home from the workplace a mean of 17 hours per week, yet 71 percent spent less than ten hours per week working from home (Statistics Canada, 2007). The amount of time people spend working at home varies, and as such we might expect that home and work experiences will also vary.

Mobile Communication

What Kind of ICTs Do Home Workers Use to Connect with Family and Peers?

Communication is also mobile. With the prevalence of mobile phones spanning socio-economic statuses, people are more connected with not only their social networks and family members, but also their workplace and organizational peers. ICTs like email have helped facilitate communication between employees and employers for those people who work at home, and have also helped self-employed individuals connect with their clients (Haythornthwaite & Wellman, 1998;). Mobile phones can provide easy and quick instrumental communication, or be used to maintain continuing connections with family and friends (Wajcman, Bittman & Brown, 2008; Ito, 2001; Wajcman, 2008).

People use many different ICTs to communicate and stay in touch; landlines, email, instant messaging, and mobile phones create a communication ensemble (Haythornthwaite, 2000; Haythornthwaite & Wellman, 1998) where people can choose what communication tool suits them depending on the situation and context (Hogan, 2008). Although some have argued that communication via email, text messages or mobile phone is too ephemeral, and too far removed from "the realities of shared space and time" (Menzies, 2005), ICTs, particularly mobile phones, can be used to sustain intimate relationships with family members and to stimulate these bonds during the workday (Baym et al., 2004; Wei & Lo, 2006; Christensen, 2009). ICTs do not replace or compensate for time spent with family members but they can supplement and help individuals adapt to the current family and household realities (Licoppe, 2004). Mobile phones can provide easy and quick instrumental interactions that help people maintain a work-life balance.

Work-Life Balance & Social Policy

With the prevalence of mobile communication and increasing number of people working at home in some capacity, the boundaries between work and home can become blurry, even if paid work at home offers people more versatility and flexibility (Schieman & Glavin, 2008; Sullivan & Lewis, 2001). Depending on how many hours they work at home, organizing the workday can be problematic for home workers, as there are also constraints of domestic life to deal with. Some workers, particularly those who spend most of their time working at home or who are self-employed, work during times that are traditionally spent with family, which can mean less time spent with them (Baines & Gelder, 2003). On the other hand, those who spend more time working at home may spend more time with their family members and experience less family conflict (Gajendran & Harrison, 2007).

What Is the Relationship between the Percentage of Time People
Work at Home and How They Integrate Paid Work, Domestic Work,
and Family Life? How Do ICTs Facilitate Work-Life Balance
and Home-Work Connectivity?

Because working at home is so flexible and diverse, conceptualizing policies that help balance work and home can be challenging. The nature of work at home has become increasingly relevant to policymakers, and the blurring of home-work boundaries continues to pose issues, but now under the rubric of "work-life balance." Canadian health and labor ministries show a growing interest in both working at home, and in how work at home impacts employee stress and people's ability to manage their work and family lives. A 2001 study by Health Canada found that three times as many Canadians experienced high stress than they did in 1991, with two-thirds of Canadians experiencing "role overload" in their attempt to negotiate work and family responsibilities, and this has had a negative impact on Canadian families; overloaded workers are less satisfied with their family life, and they often sacrifice their personal needs as a coping strategy (Health Canada, 2009). Canadian society has also been impacted by overloaded employees: fewer children, increased strain on the health care system with more visits to family physicians and hospital emergency, and more physician prescribed medication—also a strain for insurance companies (Health Canada, 2009).

Yet despite these "red flags" and significant implications to individuals and social institutions, few work-home policies exist and even fewer are actually successful. This suggests that it is vital to consider not only more flexible work arrangements, but we need to understand what that flexibility means or looks like in the context of the home; flexibility reflects the diversity of circumstances and experiences that take place when work and home overlap.

What Can We Learn from These Blurred Home-Work Spaces to Inform
the Creation of Labor Policies That Benefit Workers
and Their Families?

In light of the heightened awareness of work and family challenges and the prospects of conducting paid work in the home, we use our case study to examine home workers by comparing the percentage of time people spend working at home from minimal to full-time. Our study analyzes the interplay of households, family and relationships in order to investigate the activities and practices of home workers and to understand how they use the ICTs to mediate their home and work life.

The Connected Lives Project

Data collected for our *Connected Lives* project come from East York, a Toronto residential area located 30-45 minutes from the downtown, with a population of 112,054 people in 48,057 households (Statistics Canada, 2006a). In 2005, we employed a 32-page survey to 350 adults, which included information about how people in East York conduct paid work in the home, the amount of time spent on paid work at home and how they carry out their work tasks using ICTs (Wellman, Hogan, et al. 2006). In total, 92 survey respondents said they conducted some type of work at home. We also conducted 87 in-home interviews that included information about daily work, leisure, household relations, social networks, social routines, and ICT use. From the 92 survey respondents, 35 of the interview sample spend some percentage of their workday working at home.

Characteristics of Paid Home Workers

Who works at home and what percentage of their workday is spent working at home?

Just over half (51 percent) of the survey respondents who work at home are women, with a mean age of 42. Over two-thirds (69 percent) of the survey respondents are married or stably partnered; 64 percent of the respondents have children. Work at home survey respondents are better educated than those who do not work at home. The mean annual household income for is just over $78,000, compared to non-work at home survey participants who report a mean annual household income of $66,000. One-third (32 percent) of are employed in business, finance, and administration occupations. Just over one-quarter (27 percent) are in the social sciences, education, government service, and religion occupations, and 18 percent are in sales and service occupations (Statistics Canada, 2006b).

In today's busy world, there are many instances when people continue their workday outside the office, typically at home. However, the hours they choose to spend at home varies based on the types of tasks to be performed and other work related factors, such as peak business times for financial advisors or examination times for teachers. Other individuals who work at home may spend considerable more time working at home because of the demands of their job, or they may work primarily from home. The amount of time people spend working at home—whether a few hours a week or a full work week—will have different effects on not only their work life, but also their home lives. They are all home workers, but quite different kinds of home workers.

Categorizing the percentage of a person's work week spent at home based on typical work per week helps to differentiate the experiences of people who work at home in various capacities. We categorize three different percentages of work at home: *Full-timers* do a majority of their work at home: on average 29 hours of paid work at home per week; *Part-timers* spend 16 percent to 50 percent of their work week at home and work an average 11 hours of paid work at home per week; *Over-timers* spend 1 percent to 15 percent of their work week at home and work an average of 5 hours of paid work at home per week.

Despite the telework gurus' fascination with full-time working at home, such people comprise only one-quarter of our sample (27 percent), with the highest percentage of the respondents (47 percent) only working over-time at home (Table 12.1). Despite the stereotypical depiction of full-timers being women with children, in fact only 28 percent are women. They are also less likely to be partnered (including married) than part-timers.

Table 12.1 Profile of Home Workers

		Over-timers	Part-timers	Full-timers	Total
Demographics:					
% in Category		47%	26%	27%	100%
	n =	*43*	*24*	*25*	*92*
% of Women		41%	30%	28%	51%
	n =	*19*	*14*	*13*	*46*
Mean Age		41	42	45	43
	n =	*43*	*23*	*25*	*91*
Mean # of yrs online		7.9	8.4	8.4	8.2
	n =	*43*	*22*	*23*	*88*
% Partnered		65%	83%	63%	69%
	n =	*28*	*20*	*15*	*63*
% with Children		56%	71%	72%	64%
	n =	*24*	*17*	*18*	*59*

Table 12.1 (*continued*)

		Over-timers	Part-timers	Full-timers	Total
Mean Personal Income (CAD)		65,278	49,079	43,929	55,328
	n =	36	19	21	76
Mean Household Income (CAD)		81,293	91,964	70,132	78,025
	n =	29	14	19	62
Occupations (n = 78):					
Health		14%	21%	5%	13%
	n =	5	4	1	10
Social Sciences, Education, Government Service & Religion		27%	21%	27%	26%
	n =	10	4	6	20
Art, Culture, Recreation & Sport		0	5%	5%	3%
	n =	0	1	1	2
Natural & Applied Sciences & Related		8%	11%	9%	9%
	n =	3	2	2	7
Business, Finance & Administration		32%	32%	32%	32%
	n =	12	6	7	25
Sales & Service		19%	11%	23%	18%
	n =	7	2	5	14

Integrating Paid Work, Domestic Work and Family Life

What is the relationship between the percentage of time people work at home and how they integrate paid work, domestic work, and family life?

Domestic Work: The more time people spend working at home, the more they schedule and integrate their paid and unpaid work tasks in the home. Full-timers have a structured routine to handle the domestic responsibilities that come up during the day, and to integrate these household tasks into their workday at home. Because part-timers also do a substantial amount of work at home, it is an important part of their lives, and they must schedule around their workplace and home. By contrast, over-time workers do not tend to schedule work or have a routine for work at home tasks.

Full-timers and part-timers negotiate routines or develop coping strategies to avoid being overburdened by their work. As home becomes their workplace, many interview participants struggle with accommodating personal and family time in their hectic work schedules, realizing that they would need to erect clear boundaries between their home and work day. For part-timer Gerry, the solution is to create a temporal boundary between her workday and personal time; she takes her dog for an hour walk everyday at five o'clock, which she treats as "commuting time." When Gerry, a behavior therapist, first decided to participate in a pilot project that involved working from home, she was warned by her managers of the challenges that home workers face in balancing their home and work life. She did not expect this to be an issue for her, and has been surprised to find that it has become one. Her work is often interrupted by her daughters when they come home from school and demand her attention. Moreover, she is often kept awake at night by the thought of mistakes in her work and feels compelled to go down to her workspace and fix them.

Over-time respondents do not report the same challenges of scheduling or separating home and work as do part-timers and full-timers. They do a few specific tasks outside of formal work hours to reduce the hours they spend at work. For example, teachers find it more convenient to grade assignments and complete course preparation at home. Yet, even for over-timers, working at home presents the potential to do too much work, leading to a problematic convergence of home and work space. Several over-time workers who were interviewed cut down on the amount of work they did at home to maintain work and personal boundaries; Ruth, an over-timer who drastically reduced her home work hours, notes: "it got to the point where I was doing too much, so I think I'm in a bit of rebellion."

Doing paid work at home does not cut into the time that people have available to do domestic work. To the contrary: the more time people spend doing paid work at home, the more domestic work they do. Being at home can mean that more household chores are done throughout the day, as chores, cleaning, cooking and baking can be integrated into the home worker's day. Full-timers do the most domestic work (Figure 12.1), and many integrate household chores or childcare responsibilities into their daily home work routine. Part-timers are similar: Yvonne, a psychotherapist in the part-timer group, reports that it is preferable to blend home and work responsibilities during the workday because then "it all feels like one life." By contrast, over-timers tend to do chores in the evening or on the weekend, and they talk about their domestic responsibilities as separate from their work responsibilities.

Figure 12.1 Mean Number of Hours Spent on Domestic Work Per Week

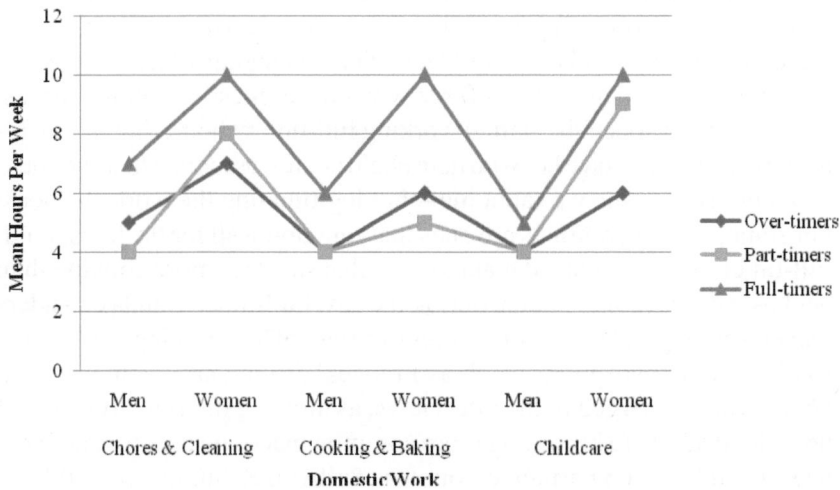

Abundant research has shown that in the gendered division of labor in the home, women do more unpaid domestic work then men (Hochschild, 1989; Luxton, 1980; Robinson & Godbey, 1997). The home lives of survey respondents also reflect this gendered division of labor. While both men and women do unpaid domestic work such as chores, cooking and cleaning, women (in all groups) still do more domestic work and integrate household chores into their work at home slightly more so than men do. Some of these couples negotiate the division of domestic and family responsibilities, notably over-timers. For example, Frances has arranged a schedule for driving her sons to school that revolves around her work at home hours: her husband takes them twice a week and she drives them the remaining days. She is responsible for picking the boys up from school all but one day a week, when she stays at work late to do paperwork and her husband picks them up.

The more time people spend working at home, the more domestic work they do. Thus, full-time women and men do more domestic work than part-time and over-time home workers. However, the relationship between the hours spent in doing paid work and domestic work is especially strong for full-time women: they spend the greatest number of hours on all household tasks.

Despite the fact that men do domestic work (and full-time men do the most), traditional gendered divisions of labor still exist. For all categories, women do more domestic work than men. Women often remain responsible for domestic work and childcare after they begin working at home

(Sullivan & Lewis, 2001; Sullivan & Smithson, 2007), which can create higher levels of work-family conflict for women than men (Ahuja, 2002). Gendered divisions of labor and a lack of understanding of home-work by other household members reinforce women's domestic responsibilities. Olivia notes that once she started working full-time at home, her family reduced the help they gave her with household chores, assuming that she could do them because she was at home—not legitimating the work she does. For some, existing routines are ones that function well for the household: full-timer Theresa thinks it makes sense that she does more cooking than her husband since she *is* home during the day. Full-timer Wanda considers dinner planning and household chores to be part of her morning work-home routine. Some exceptions, such as Frances' driving arrangements with her husband, emerged in the interviews, as more egalitarian divisions of household labor are becoming more common place and are more actively negotiated between partnered couples (Sullivan & Smithson, 2007).

Some respondents enjoy the convenience of integrating household tasks into their workday. Full-timer Beth puts in a load of laundry or empties the dishwasher while she prints documents or downloads files for work. On the other hand, Olivia feels that she has to avoid getting preoccupied with domestic chores and focus on her work tasks. Part-time behavior therapist Gerry says that she does a disproportionate amount of the household chores, but is balanced with her being more involved in the day to day activities of her daughters when she is working at home than her husband.

Childcare: In addition to household chores, the more time people spend working at home, the more childcare they do. Full-time parents spend the most time on childcare, but the most drastic difference between women and men is in the amount of childcare they do. While both women and men who do paid work from home do childcare, it remains primarily women's responsibility. Women spend twice as many hours doing childcare than men, with full-time women spending the most time taking care of children.

Some full-time women try to balance childcare responsibilities and household chores. Theresa, a policy analyst in the full-time group, has done free-lance work from home since the birth of her third child. Her children's schedule influences the scheduling of paid and unpaid work: her three and a half year old twin daughters go to preschool three days a week while she cares for her infant son. Family schedules strongly influence how full-timers arrange their workday. Some home workers deliberately organize work hours around their children's school schedule or their spouse's work schedule. For example, Sean, a full-time guitar maker helps prepare his son for school and his wife and work on weekday mornings. He picks up their son from school in the afternoon and gets dinner ready. He does much of his paid work on the weekends when his wife can watch their son.

By contrast, James, a self-employed Web designer and single father of a seven-year-old boy, notes that it is important to him that his work at home does not interfere with spending time with his son. He works 50 to 60 hours a week at home during the week, and he devotes his weekends to his son. He also helps him with homework in the evening, so most of his work gets done on weekdays before three in the afternoon and after his son has gone to bed. Beverly, a full-timer who runs a home daycare, feels that it is comforting to her son and live-at-home elderly father that she is easily accessible throughout the day.

The more time people spend working at home, the more time they spend with their children (in addition to giving them care). Full-timers spend the most time with their children, including watching television (Figure 12.2). Full-timers also spend almost twice as many hours per week using the Internet with their children than part-timers and over-timers. This is perhaps partially due to their mostly being in white-collar occupations. These differences remain when we look only at women. Women in all groups spend more time overall with their children than men.

Figure 12.2 Mean Hours Per Week Spent with Children

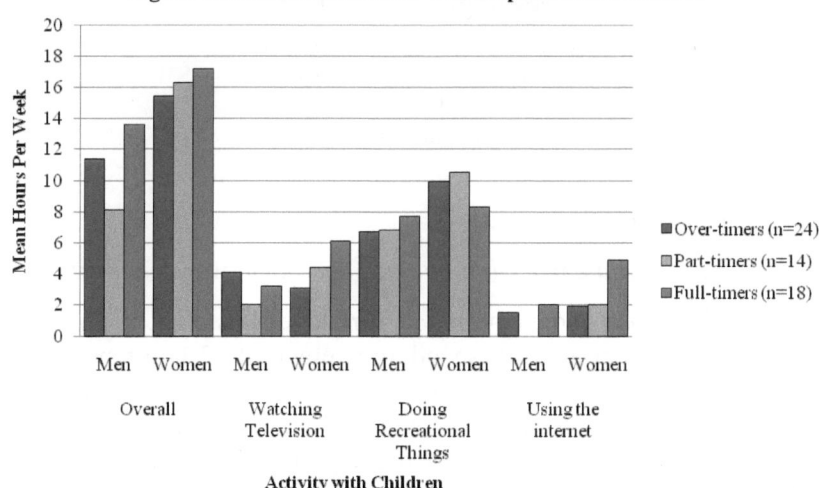

Time with Partners: The more time people spend working at home, the more time they spend with their partners: full-timers spend the most time overall (not including time spent sleeping) with their partners, slightly more than part-timers and over-timers (Figure 12.3). Full-timers also watch more television with partners than part-timers do, but less than over-timers. Full-timers spend the least number of hours doing recreational things with their spouse compared to part-timers and over-timers. By contrast to television watching, full-timers spend slightly more time on the Internet with their partners, a pattern similar to time spent online with children.

Figure 12.3 Mean Hours Per Week Spent with Partner

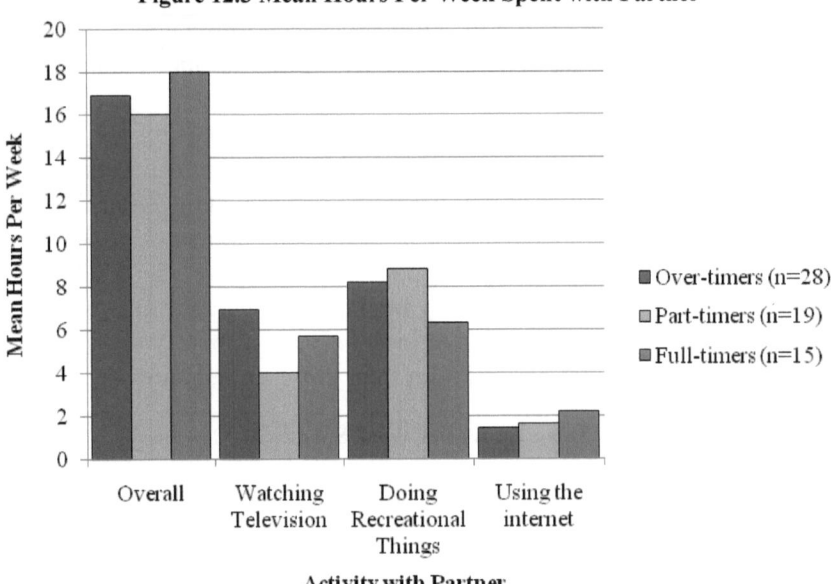

Using ICTs to Stay Connected

How do ICTs facilitate work-life balance and home-work connectivity?

ICTs at Home and at the Workplace: Employed East York respondents who use the Internet at work spend an average of 13 hours per week using the Internet, and they spend on average eight hours of this time specifically on work related tasks. Clearly, workers are incorporating non-work related tasks throughout the day, including communicating with their partners, children, friends and relatives.

Although ICTs are not needed to carry out every type of work at home, the salience of the Internet in the typical twenty-first century worker's day is evident in all three types of home workers: no matter how many hours people spend working at home, most of them use the Internet in some capacity for work.

Despite the increasing prominence of Internet-based communication, the traditional landline telephone is the most widely used communication device for people who work at home. As might be expected, full-time workers use their landline more often than part-timers or over-timers.

Emailing has become so integrated into contemporary office life, that all types of workers—full, part, and over-timers—use it extensively. Wherever people do most of their work—whether home or at the workplace—is the place where they do most of their emailing. Primary work location determines where emails are sent from most often. If one works more at home, more emails are sent from home and if one works more at the workplace, the more emails are sent from the workplace.

As emails can be sent from any location and can be sent at any hour of the day, they might seem to be the ideal task to be relocated to the home by the over-time worker. This is not the case. Over-time workers prefer to conduct email conversations from their main workplace. The interview data suggest that many of the over-time workers in our sample made a conscious decision to not be reachable by email at home, recognizing the potentially invasive nature of email and mobile devices.

Communication with Partners: Despite the challenges of finding face-to-face time with their families, home workers do stay connected with their spouses and children using ICTs such as landlines, mobile phones and email. The more time people spend working at home, the more contact they have with their partners via ICTs throughout the day—no matter where they are or what they are doing (almost all partnered home workers have employed spouses). Full-timers connect with their partners via ICTs slightly more often than part-timers and over-timers (Figure 12.4). Yet each work group has a communication device they tend to prefer, with the type of device reflecting their daily mobility. Because full-timers are most often at home, they often use stationary landline phones to connect with their partners. By contrast, part-timers tend to use mobile phones, perhaps because communication can be from anywhere throughout the day, an important affordance for those who are always on the go. Over-timers, away from home more than the others, send more emails to their spouses than do the other groups. There are also gender differences in how ICTs are used to connect with spouses: men communicate slightly more often with their spouses than women across all types of communication media.

Figure 12.4 ICT Communication with Partner

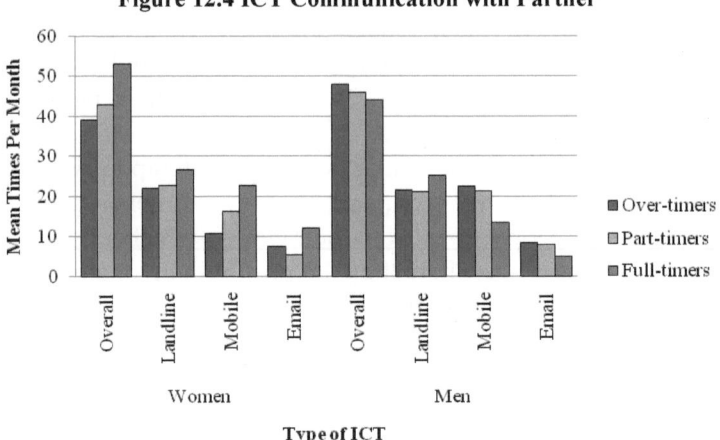

Communication with Children: The more time people spend working at home, the less they use email and mobile phones to contact
their children throughout the day (Figure 12.5). However, the use
of landlines to contact children is similar across all home worker
groups. Because full-timers spend the most number of hours at home,
and the most time with their children face-to-face, it is not surprising
that they communicate less with them by email or mobile phone than
part-timers and over-timers. Moreover, given that women in all home
worker groups spend more time with children (leisure and childcare)
than men, it makes sense that women use ICTs more often than men
to connect with their children.

Figure 12.5 ICT Communication with Children

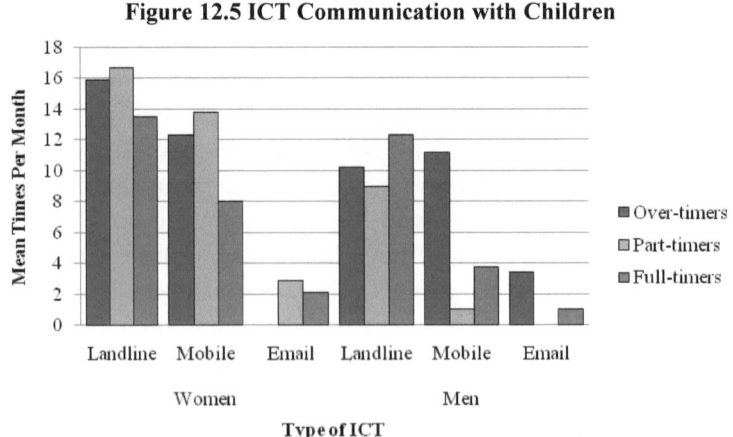

Communication with Friends and Relatives: Full-timers spend more time at home than other home-workers, and use their home Internet to send more email messages (Figure 12.6). Not surprisingly full-timers send more work-related emails than part-timers and over-timers who more often go physically into their offices. Full-timers are also somewhat more likely to send emails to their household members, friends and relatives. Women email only slightly more than men do across all work groups. By contrast, workers who spend more time at work outside of the home, especially over-timers, do most of their emailing from the workplace (Figure 12.7).

Figure 12.6 Mean Number of Emails Sent From Home Per Week

$$*p = <.05$$

Figure 12.7 Mean Number of Emails Sent from Workplace Per Week

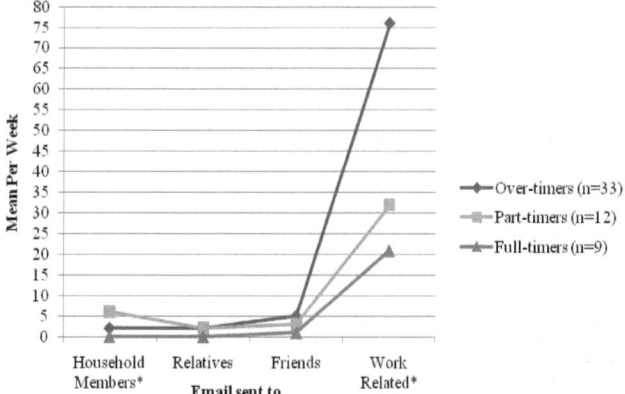

$$*p = <.05$$

Conclusions

Gurus, managers and ordinary people have recognized the potential of ICTs to facilitate working at home and their family life. The mobility and ubiquity of ICTs facilitate, complement and enhance communication between home workers and family members throughout the day. The pervasiveness of the ICTs in people's lives is unquestionable; it is integrated into social and leisure activities, educational pursuits and work lives. The affordance of convenience and ease, increasing mobility, portability and accessibility to ICTs continues to complement how people carry out their days. At the same time, major stresses at home emerge from the interplay between household chores, raising children and doing paid work.

In part, this is because home workers vary so much. While the eyes of gurus is on those working full-time at home, in practice, most work at home part-time and over-time. There are important variations in how people in these different categories handle their paid work, domestic work and familial relations.

Full-timers understandably have the most structured and the most scheduled arrangements. They also spend the most time doing domestic chores and raising their children. To some extent, this is understandable because they are at home the most. But our findings also show that the double load is greatest for them of doing domestic work and paid work. And the load is greater for women. Full-time, part-time and over-time they consistently do the most domestic work. In short, the double load is greatest for women and working full-time at home.

The variation in how people manage home-work situations becomes clearer when we look substantively at full, part and over-timers. Full-timers have blurrier boundaries between work and domestic life than part-timers and over-timers. Full-timers establish routines for their paid and unpaid work, yet their tasks overlap more than other home workers. While some full-timers carefully segregate work time from domestic and family time, most disperse their tasks throughout the day. This contrasts with over-timers who have much more separation between work and home. Likewise, while full-timers do more cooking, cleaning and childcare than part-timers and over-timers, they also spend more time with their children and spouses. Full-timers connect with their partners via landlines and ICTs most often throughout the day, but least often with their children who they see before and after school. While all home workers use ICTs to stay in touch with household members, their mode of contact varies: full-timers use landlines, part-timers use mobile phones, and over-timers use email.

Policy Implications for Working at Home

What can we learn from these blurred home-work spaces to inform the creation of labor policies that benefit workers and their families?

The varied experiences between different types of home-workers and the different ways of utilizing ICTs show the need for flexible labor policies. Full-timers, part-timers and over-timers have different ways of integrating paid work, domestic work and family life, and negotiating the work-life balance. The differences between them are more than their hours of work: they have qualitatively different home-work routines. The more time people spend working at home, the more integrated, embedded, and blurred their work and family life becomes. Most importantly, the more time they work at home, the more time they spend with family members—for better or worse. Consequently, we encourage organizations and governments to support home-work arrangements in varying capacities, and to consult individuals about their own needs and the needs of their household members.

Significant issues become less visible in the home and need to be considered, such as employee health, overtime regulation, insurance, ergonomic furniture, and technical support. There is a need to ensure that unionized workers are covered by their collective agreements while working outside the workplace, that health and safety standards continue to be adhered to, and that workers are not isolated from union representation or from training and promotion opportunities. In Canada, only a few collective agreements with crown corporations and public sector offices extend to work at home arrangements. The few North American regulatory labor policies that exist focus on manual piecework (e.g., Johnson 1982), and do not address the era of mobile and networked work. The number of people who continue their work day after hours has increased in recent decades, yet there are no clear policies for white-collar workers let alone work at home policies that can help people deal effectively with role overload (Health Canada, 2009).

Workplace policies stem from workplace cultures, and work culture is an important impetus of role overload; when organizations are supportive of work-life balance, role overload is greatly reduced

(Health Canada, 2009). Our analysis shows that flexibility is a key to labor policy initiatives and reform, and as such we call for:

1. Flexible work locations: policies to help people deal with role over-load and physical, mental or emotional fatigue often associated with employee absenteeism;

2. Flexible work schedules: encourage organizations and governments to support work at home arrangements in varying capacities;

3. Flexible work provisions: consult workers about their own needs and the needs of their family members.

Policies need to be fluid. Because life circumstances change, ongoing interaction between employer and employee about family needs is needed. Organizations must realize that flexible work-life policies will benefit them: employees will be more committed to the organization, they will have less work stress, and they will be more satisfied with their jobs and well-being (Health Canada 2009).

Acknowledgments

Our research has been supported by the Social Science and Humanities Research Council of Canada, and the Intel Corporation's People and Practices unit. We are gratefully for the advice of NetLabbers Kristen Berg, Jeffrey Boase, Bernie Hogan, Jennifer Kayahara, Julia Madej, Sinye Tang and our transcribers and coders.

References

Ahuja, M. K. (2002). Women in the information technology profession: A literature review, synthesis and research agenda. *European Journal of Information Systems,* 11 (1), 20-34.

Baines, Susan & Ulrike Gelder. (2003). What is family friendly about the workplace in the home? The case of self employed parents and their children. *New Technology, Work and Employment,* 18 (3), 223-234.

Baym, Nancy, Zhang, Yan Bing & Mei-Chen Lin. (2004). Social Interactions across Media: Interpersonal Communication on the Internet, Face-to-Face, and the Telephone. *New Media & Society,* 6(3), 299-318.

Cairncross, Frances. (1997). *The death of distance: How the communications revolution will change our lives.* Boston: Harvard Business School Press.

Christensen, Toke Haunstrup. (2009). 'Connected presence' in distributed family life. *New Media & Society,* 11(3), 433-451.

Dimitrova, Dima. (2003). Controlling telework: Supervision and flexibility revisited. *New Technologies, Work and Employment,* 18 (3), 181-195.

Gajendran, Ravi & David Harrison. (2007). The good, the bad, and the unknown about telecommuting: Meta-analysis of psychological mediators and individual conse-quences. *Journal of Applied Psychology,* 92 (6), 1524-1541.

Gordon, Gill & Marsha Kelly. (1986). *Telecommuting: How to make it work for you & your company.* Englewood Cliffs, NJ: Prentice Hall.

Gordon, Gill. (1987). The dilemma of telework: Technology vs. tradition. Paper pre-sented at the Telework—Present and Future Development of a New Form of Work Organization, Bonn, Germany.

Haythornthwaite, Caroline. (2000). Exploring Multiplexity: Social Network Structures in a Computer Supported Distance Learning Class. Working Paper. University of Illinois at Urbana-Champaign: Graduate School of Library and Information Science.

Haythornthwaite, Caroline & Barry Wellman. (1998). Work, Friendship, and Media Use for Information Exchange in a Networked Organization. *Journal of the American Society for Information Science,* 49(12), 1101-1114.

Health Canada. (2009). Work-life conflict in Canada in the new millennium: Key findings and recommendations from the 2001 national work-life conflict study. Retrieved February 25, 2010 from: http://www.hc-sc.gc.ca/ewh-semt/pubs/occup-travail/balancing_six-equilibre_six/chap3-eng.php.

Hochschild, Arlie. (1989). *The second shift: Working parents and the revolution at home.* New York: Viking Penguin.

Hogan, Bernie. (2008). Analyzing Social Networks via the Internet. In Nigel Fielding, Raymond Lee & Grant Blank (Eds), *The Handbook of Online Research Methods* (pp. 141-160). Thousand Oaks, CA: Sage.

Hogan, Bernie, Carrasco Juan-Antonio & Barry Wellman. (2007). Visualizing Personal Networks: Working with Participant-Aided Sociograms. *Field Methods,* 19, 116-144.

Ito, Mimi. (2001). Mobile Phones, Japanese Youth, and the Re-Placement of Social Contact. *Proceedings of Annual Meeting for the Society for the Social Studies of Science (*pg 10). Cambridge, MA.

Johnson, Laura. (1982). *The seam allowance: Industrial home sewing in Canada.* Toronto: Women's Educational Press.

Licoppe, Christian. (2004). 'Connected presence': the emergence of a new repertoire for managing social relationships in a changing communications technoscape. *Environment and planning D: Society and space,* 22, 135-156.

Luxton, Meg. (1980). *More than a labour of love: Three generations of women's work in the home.* Toronto: The Women's Press.

Menzies, Heather. (2005) *No time – Stress and the Crisis of Modern Life.* Vancouver: Douglas & McIntyre.

Schieman, Scott & Paul Glavin. (2008). Trouble at the border? Gender, flexibility at work, and the work- home interface. *Social Problems,* 55(4), 590-611.

Statistics Canada. (2006a). *2006 Census: Families, marital status, households and dwelling characteristics.* Retrieved February 25, 2010 from: http://www.statcan.gc.ca/daily-quotidien/070912/dq070912a-eng.htm.

Statistics Canada. (2006b). *National occupational classification for statistics (NOC-S).*

Retrieved February 25, 2010 from: http://dsp-psd.pwgsc.gc.ca/collection_2007/statcan/12-583-X/12-583-XIE2007001.pdf.

Statistics Canada. (2007). Working at home: An update. *Perspectives on Labour and Income.* 8(6). Retrieved February 25, 2010 from: http://www.statcan.gc.ca/pub/75-001-x/10607/9973-eng.htm.

Sullivan, Cath & Susan Lewis. (2001). Home based telework, gender, and the synchronization of work and family: Perspectives of teleworkers and their co-residents. *Gender, Work and Organization,* 8(2), 123-145.

Sullivan, Cath & Janet Smithson. (2007). Perspectives of homeworkers and their partners on working flexibility and gender equity. *International Journal of Human Resource Management,* 18 (3), 448-461

Wajcman, Judy. (2008). Life in the fast lane? Towards a sociology of technology and time
British Journal of Sociology, 59(1), 59-70.

Wajcman, Judy, Bittman, Michael & Judith Brown. (2008). Families without Borders: Mobile Phones, Connectedness and Work-Home Divisions. *Sociology*, 42 (4), 635-652.

Wei, Ran & Lo, Ven-Hwei. (2006). Staying connected while on the move: Cell phone use and social connectedness. *New Media & Society*, 8(1), pp. 52-73.

Wellman, Barry & Bernie Hogan, with Kristen Berg, Jeffrey Boase, Juan-Antonio Carrasco, Rochelle Côté, Jennifer Kayahara, Tracy L.M. Kennedy & Phouc Tran (2006). Connected Lives: The Project. In Patrick Purcell (ed.) *Networked Neighbourhoods* (pp. 157-211). Guildford, UK: Springer.

13

M-enabled Learning: The Mobile Phone's Contribution to Education

Lourdes M. Portus

As education adapts and copes with modern technologies, students and professors go "high-tech," using the computer and the Internet in obtaining knowledge and information. There is, however, a dearth of evidence on how the emergent mobile phone itself actually contributes to learning. Thus, developing learning activities that capitalize on the mobile phone's unique features of size, portability and widespread availability is a great challenge for professors and academic institutions alike.

The mobile phone is a modern ICT phenomenon inflicting all facets of life in society, including the educational system. As ICTs are integrated and harnessed to improve quality of education, evidence on the mobile phone's contribution to education is hereby problematized.

Reminiscent of Katz's (2005) study on mobile phones in educational settings, some professors prohibit the use of the mobile phone, categorically stating, for instance, in their syllabi: "to turn off mobile phones or put them on silent mode during class." A penalty of sorts is imposed on violators, e.g., submit a think paper, treat the class to a scoop of ice cream, or serve snacks, etc.

This prohibition implies a contentious regard for the mobile phone as a distraction, and one that is completely disconnected to classroom learning. Professors would feel insulted, or taken for granted, whenever students step out of the room to answer a phone call or text message, or whenever students withdraw from class discussions in order to do so. Campbell's (2002) study reinforces the problems and interruptions that the mobile phone's ringing causes in the classroom and it supports the formulation of policies that would effectively ban the mobile phone from the classroom.

Mobile phone-enabled learning refers to the mobile phone's use allowing students and professors to connect with each other, primarily for academic purposes, as well as with classmates, family members and friends. It also means that, given a 100 percent-wireless signal, students and professors, whether in campus or elsewhere, can share relevant class and university information (Benson 2008).

I use the phrase, "m-enabled learning" instead of "m-learning," because of the presumed facilitating, albeit indirect, contribution of the mobile phone to learning. The presumption stems from the generally known limitations of the mobile phone, such as, its small display screen and keypad, 160-character capacity, one-on-one interaction and short battery life. The degree of literacy of the user even heightens the mobile phone's limitations (Chipchase 2008).

Considering the students' middle-class status and, subsequently, their inability to acquire the prohibitive iPhone and BlackBerry, I would assume that the student-informants couldn't use the mobile phone as a discussion board or a chat venue.

Meanwhile, I define "enablement" as the mobile phone's ability to facilitate access to information for both students and professors (adapted from Alfonso, 1999). As an "enabler," the mobile phone creates numerous possibilities for communication and knowledge-sharing.

Hence, this study presents significant data regarding the student-professor interaction via the mobile phone—the students and professors being the principal stakeholders in the learning process.

M-enabled learning builds on an obvious fact of life in campus. I have yet to encounter a college student from the University of the Philippines (UP) who does *not* possess a mobile phone. I wonder whether or not, apart from intimate social interactions, students also use the mobile phone as part of their paraphernalia for their educational endeavors.

In what ways and contexts has the mobile phone aided students in effectively coping with course requirements and accomplishing academic goals? With the mobile phone's technical limitations, how can it then contribute to the students' learning or education?

Research Problem and Objectives

How does the mobile phone contribute to students' learning? Does it enable students to acquire knowledge, or facilitate the students' learning process? Or, does it disrupt the students' learning, and consequently, their academic pursuits? Specifically, this study aims to:

1. Describe the practices and manner in which students use the mobile phone, negotiate its use and cope with its limitations;
2. Find out the extent of enhancement or disruption that the mobile phone brings to the learning process;
3. Assess the student-professor interaction via the mobile phone in terms of nature and frequency, content, effect and policies; and
4. Describe and assess the enabling and socio-cultural process that students undergo in using the mobile phone for academic purposes.

Research Methods

I employed mainly qualitative research methods, drawing data from the textual analysis of the diaries of 25 selected students and from interviews with 25 selected professors, who use the mobile phone for both personal and academic purposes. Afterwards, I derived insights and implications from the diaries, which reflect the students' individual experiences during the period, June to September 2008, as well as from the results of the interviews with professors.

The study's student-informants, all communication research majors, were my former students in a qualitative research course. Following specific instructions, they recorded their daily experiences in using the mobile phone vis-à-vis their academic pursuits and included their reflections on the mobile phone's importance to themselves and their studies.

Trained and exposed in field research, each student-informant interviewed one professor, who uses a mobile phone in teaching. Thus, under my direct supervision, the students interviewed 25 professors from the University of the Philippines.

Also, drawing information from secondary sources, I reviewed pertinent articles from UNESCO (United Nations Educational, Scientific and Cultural Organization) publications, specifically that of Braid and Tuazon's (2007). The November 2008 ICT (Information Communication Technology) and Education Workshop, which the UNESCO sponsored, and which I attended, have stimulated relevant insights as well.

Related Literature the (mobile phone)

Most of the literature on the mobile phone explores the latter's contribution in providing information, networking, commerce, self-expression, romantic and filial relationships, love and intimacy, environment and entertainment. Also, there exists literature on the mobile phone's impact on learning. This includes (a) the Katz et al. (2005) study on mobile phone use in a classroom setting that finds some similarities in this study and which is now given local Filipino color; (b) Campbell's (2002) "Mobile Phones in College Classrooms," which deals with ringing, cheating and policies; and c) Quirante's (2006) findings on using abbreviations in text messaging and its deleterious effects on the English language, sentence construction and grammar.

Resonating with Campbell's (2002) study, Portus, Quirante et al.'s (2007) study illustrates how the mobile phone promotes irresponsibility; it, however, fails to establish how the mobile phone contributes to learning.

Overall, the mobile phone's role in facilitating learning, particularly in the context of Filipino students, has not been fully explored.

Text2Teach

A significant development in the mobile phone's role in contributing to quality education is "Bridge IT," an international collaborative alliance of Nokia, International Youth Foundation, Pearson and the United Nations Development Program (Catangay 2009: A11). It *has* spearheaded the *"Text2Teach"* Program—a consortium of the Philippine Government's Department of Education, Ayala Foundation (Globe, Telecom, Nokia) and SEAMEO INNOTECH (Southeast Asian Ministers Education Organization and Regional Center for Educational Innovation and Technology). The program aims to "narrow the educational divide between nations by improving the teaching of basic education in developing countries using high-speed, wireless digital connection." (Dumlao 2009: B24)

Text2Teach or T2T is a text-message-based program that uses the mobile phone because almost everyone has it and knows how to use it. The mobile phone is cheap and can reach the farthest islands in the Philippines (Dumlao 2009: B24). It is intended for Grades V and VI students who are provided with multimedia packages that make English, science and math meaningful to them.

There are two phases *in using T2T*. The first phase involves using the mobile phone to order video clips, which are delivered via satellite, stored in the media master and viewed on *television*.

The second phase uses the mobile phone as storage of educational videos. The mobile phone is then plugged to a TV set where the videos are played. A Text2Teach professor sends a code that represents a certain video to a Globe (a local service provider) number. The network transmits the text to the service provider, which then converts the text into ASCII (American Standard Code for Information Interchange) to be sent through e-mail. The service provider sends the ordered digitized content video to the satellite network operator. The digitized content video is sent off to the satellite and is received by the satellite dish to be downloaded immediately to the Nokia Media Master, provided the power is turned on always. Once it is recorded in the Media Master, the professor may then view the video from the TV set in the classroom. About a hundred video clips in science were made available through the Text2Teach system.

According to some evaluation reports, the T2T program has resulted in reduced absenteeism among students, improved grades, more frequent interaction with professors and among classmates, and a more interesting classroom atmosphere.

While the T2T program promises to make learning accessible through the mobile phone, I think that the requirements in acquiring mobile phones with Wi-Fi capability to view educational video clips are still prohibitive in a Third World country like the Philippines.

Another requirement—to connect to a television set—would need the Nokia Media Master system set-up, which is not widely available.

The T2T program is in its infancy and available only to high school students. It remains vague on how it could benefit College students as well as out-of-school youth. Besides, the use of the T2T's gadgets, supplied by pre-identified companies, Nokia and Globe, makes one speculate on their ulterior motives.

Findings and Discussion

Negotiating Practices on Mobile Phone Use for Academic Activities

To cope with the limitations of the mobile phone, students usually switch to the Internet. Students use the "Yahoo! Messenger," instead of the mobile phone, for a more interactive discussion on their class work. Doing conferences online, in lieu of face-to-face class discussions, saves time, money and effort. Students convey messages concerning class lessons through the Internet's "Yahoo! Groups" since it is technically impossible to use the mobile phone's limited set of characters. Also, messages become vague and difficult to read when words are abbreviated, vowels omitted and letters combined with numbers.

But, despite the abovementioned limitations, the student-informants cite various contexts wherein the mobile phone displays more advantages over the Yahoo! Groups. For one, students and their mobile phones are inseparable, wherever they go. For another, text messages sent are instantly received. Of course, using the Yahoo! Groups system requires that users should first gain access to the Internet.

Moreover, students who are dorm residents would have to venture out to visit the local Internet café. These students regard the excursions as inconvenient and bothersome, hindering them from checking their e-mails. Pinky, in her diary, proves this:

> Even if it's possible to use the Internet, YM, and landline phone, the mobile phone is still more accessible and convenient to use because it is quite easy to use and students always have their hands on it. Hence, they are immediately informed in case their professor has something to tell them. Today's students seldom let go of their mobile phones, which is why it's easier to contact them through text messages or to call them through the mobile phone.

Pinky further writes that, if she wanted to use the Yahoo! Messenger, she would have to go out (looking) for a computer. But, with a mobile phone, conversations can be done anywhere, even while one is busy doing something else. Furthermore, the mobile phone provides a ready connection whenever one's classmates lose the Internet connection, especially during power outages, technical malfunctions, and poor signals.

Coping with "No-Mobile-Phone Policy" inside the Classroom

Students usually set their mobile phones to silent mode inside the classroom, or use the phone's vibration feature, lest they miss a text message. Even with the mobile phones in their pockets, students would know whether or not there were emergency situations or incoming calls.

Some students abruptly go out of the classroom to read their text messages, without waiting for the end of the class. Others secretly look at their concealed mobile phones to read the text messages.

Quotes or forwards are usually not read, but deleted immediately, to allow one to re-focus on the ongoing class discussion. Reading the entire text message is seldom done while one is in class, though, for fear of being noticed by the professor.

Message Relay

It would be impractical for professors to send text messages to all their students. The established practice is to appoint a "pass-on leader," or relay person, whowould be responsible for "texting" everyone else in class.

It should be noted that the responsibility of disseminating text messages, even at a rate of one peso (approx., US$ 0.02) per text message would mean substantial costs for the class leader—who's usually a jobless, non-earning student—especially if he/she belonged to a big class of 20 to 30 students. If he /she were to send two to three messages per day to the entire class, his/her daily allowance (courtesy of his parents) would surely be depleted in no time. Therefore, he/she would have to resort to the system of pass-on messages and request two or three other classmates to relay the message.

Uplifting or Upsetting Text Messages

Text messages among classmates either uplift spirits, or get into their nerves, depending on the type of text messages being traded. Text messages normally ease one's feelings when these are reassuring or provide solutions to class problems. Good text messages usually come from serious and conscientious students, who initiate discussions, who think about getting good grades, and who are always present during coordination and group meetings. There are classmates who encourage and influence their peers in contributing to class projects. They solicit information, show concern for paper submission deadlines, and suggest ideas.

Meanwhile, exasperating and upsetting text messages usually announce: "...I cannot come," "...I am busy" or, "...I have other commitments." These usually originate from the parasitic type of classmates, who depend on others for good grades and paper submissions. They rely on their groupmates to finish the paper, while they stray away to join other extra-curricular activities. They give all types of excuses or alibi whenever they cannot perform their tasks. And, in all these cases, the mobile phone has metaphorically become an "unwitting" tool to justify the students' non-performance.

Disturbing text messages turn out to be barriers to effective communication. Despite the highly touted, handy tool that the mobile phone has become, enabling people to communicate wherever and whenever they wish to, there would be instances, however, when people simply, or technically, are barred from communicating and/or cannot be reached.

The ability to communicate is hinged on the financial capability of middle class students whose mobile phones are usually pre-paid, meaning that they have to load (or pay) up before they can use their phones. However, there are times when they would inadvertently exhaust their load, thereby impairing their ability to contact classmates. In such a dire situation, they would immediately request a friend or relative for a "pasa-load," a system wherein a user with an abundant load in his SIM card transfers a small, but sufficient, amount of load to the user-in-need. Or, as a last resort, they would use a landline phone.

The Many Facets of the Mobile Phone in the Academia

Announcer and Notifier

The mobile phone comes in handy as the "announcer" to students for important class- or lesson-related messages being conveyed via the Yahoo! Group. It assumes the roles of "notifier" and "reminder"—informing classmates that one is online, or that one may be late, or cannot make it, to a chat session. It also reminds everyone about an online conference, or a face-to-face meeting.

A student writes in her diary:

> Our main goal is to meet up online. But, we cannot do this without the mobile phone. Although we may be in different locations at a particular time, the mobile phone gets us together by notifying and/or reminding us who is/are already online.

Coordinator and Organizer

Students repeatedly write in their diaries that the mobile phone's relevant features of portability, accessibility, speed and minimal cost, all combine to facilitate a smooth learning environment. Messages via the mobile phone deal with class lessons; friendly reminders; reporting about parts of a research paper completed for class research projects; requests to check work; and, queries or information on what needs to be done.

Since the local practice among students is to divide the tasks in a class event, exhibit, symposium, field trip, writing a group research paper and other extra-curricular activities, the best means to coordinate such activities would be through the mobile phone.

Venue for Negotiation

The mobile phone becomes a venue for negotiation between professor and learner. A professor notes students' increasing use of the mobile phone to ask him (the professor) to reconsider deadlines, project extensions, etc. Hence, the mobile phone makes him uncomfortable. But, over-all, it improves or enhances the functional relationship between professor and student.

A student stresses:

> I may not call him, if without the use of phone, and if it wasn't through the phone, I may not interview him, and if I haven't interviewed him, I may not have done the requirement for the course, and if I haven't done the requirements, then I may not pass, or can possibly fail the course. That really makes me think it is useful.

A Stressor, a Pressure

Some students describe the mobile phone as "a stressor, a pressure." The various ringtones, by which the source may be identified, bring along pressure, stress, and a fast heartbeat. These would likely jolt their recipients, especially when the latter have pending tasks. The ringtones may indicate someone following up on another's attendance, or the submission of a paper, or a reply to a question. They keep one up-to-speed with matters left undone by classmates.

The pressure mounts when the ringtones come in frequent intervals. But, then, there are ways to temper the pressure. As one views the text messages and sees Smiley emoticons, or the words, "ha! ha!" or "he! he!" or "hihi!," and even greetings, like "gud pm" or "gud am," the pressure as well as the anticipation of incoming serious text messages disappear.

A student writes in her diary:

Rein's text message claims the need for us to talk on a very serious note. Good thing his next message started with "haha!" and so the uneasiness disappeared. It's fun to know that a single word or even emoticons somehow reflect your mood, in turn, shaping others as well.

Students' testimonies in their diaries disclose that text messages from professors add to the pressure, particularly when they ask whether or not students have completed their assignments, or when they give students another assignment.

A diary entry reveals:

Wheww! We should really take courage when sending text messages to professors and breaking some sort of bad news to everyone else. *Professors' text messages create deep impact among students* (emphasis mine). It's just a text message but see how deep its impact is on me that I am actually bothered. This is not enough to prove that I should somehow pay attention to this matter.

Extension beyond the Classroom

As McLuhan predicted, the medium or technology, and in this case, the mobile phone, has become an extension of the human body and of learning activities. Both students and professors can hardly proceed with their daily routine without the mobile phone.

Students have more concerns, issues, suggestions or problems that cannot be normally discussed in face-to-face (f2f) mode, but, now, these can be expressed openly. Some students who are intimidated, or have difficulties talking f2f with their professors, can now breathe easy and reduce their apprehension and anxiety. They can send text messages to their professors—to ask for a postponement of a deadline; to be excused from attending class; to explain one's absence, or to inquire about an appointment, interview or consultation. These student concerns can now be resolved through the mobile phone.

Bridging the relationship between students and professors, the mobile phone prompts a closer relationship that can even lead to non-academic interests. Exchanges with the students happen even after class hours and do not stop within the four walls of the classroom. A professor, who may want to know more about students needing more attention, does so through his/her mobile phone.

On the professors' part, disseminating information to the students by itself helps them relate, and become attuned, to the students' personalities, needs, likes and dislikes, jokes, etc. As professors reach out to them and learn about the student's lives, they themselves exude youthful feelings.

With the mobile phone, the classroom walls fade away as professors and students exchange course-related text messages. Students and professors then obscure the distinction between their private and professional lives.

A student attests to this in her diary:

> Sometimes there seemed to be a blurring of one's private and professional lives. A long time ago, when mobile phones were not yet a trend, I think people were able to compartmentalize their private and professional lives or for our case live as family members and students. Upon leaving school or the office, one turns his/her private life (e.g., life as a daughter, mother, father, etc.).

In these times when mobile phones are used by almost everybody, such a separation of one's professional life from his/her private life seems to be impossible. Wherever you go, when something urgent pops up, there seems to be a need to respond to it because of the mobile phone's very existence. There is an expectation or urge for one to immediately answer text messages or calls.

An Academic Help

Text exchanges regarding academic matters occur, not only among classmates, but with family members as well. Being bombarded with text messages from her younger brother, who's in high school, "Ate" (elder sister), writes in her diary:

> He was always asking for academic help from me, asking if his grammar, parallelism, use of words, and subject-verb agreement are correct. I'm glad that, through text messaging, I can be of help to him in his homework and academics, aside from

psychological and emotional needs. At least, the geographical distance between us is bridged by the use of mobile phones.

Venue for Discourse

Discourses do not only occur inside the classroom. When one possesses a mobile phone, a virtual venue for discussion exists. Mobile phones become sites for the free flow of ideas among classmates and their professors. A student writes in her diary:

> However, there are few instances where I was truly moved by a line in a book that I read, a movie that I've watched, or song that I've heard that I composed a text message and sent it to close friends. I agree with Sir Xiao as he says that text messaging provides a venue for a free-flowing exchange of ideas. This is especially true for me during the above-mentioned instances.

Professor-Student Interaction via the Mobile Phone

Records of interviews with professors reveal various reasons why professors send text messages to students, including, but not necessarily limited to, emergencies, suspension of classes, directions concerning students' examinations, assignments, exam dates, schedule of activities, appointments or consultation schedules, student's status in class, reminders or announcements, etc.

Professors use the mobile phone because of the convenient and fast delivery of messages. Students may initially tend to exult upon learning from the relay person that their class has been suspended. But, they would later feel disappointment creeping in after discovering that the research or reading assignment has proven to be a more difficult task than attending the suspended class itself.

Professor's Text: A Status Symbol

A professor's text message is taken to mean a number of things: that the recipient may be a favorite student, or that he or she has a certain stature in class, or that he/she may be chosen as a relay person. Some students become envious of their classmate, who gets a text message from a professor.

Also, when a professor sends a text message to a student, the student interprets this as a sign of trust, that he is perceived to be a mature and responsible person. In fact, some students would feel privileged and brag about receiving a text from a professor.

Receiving a text from a professor may bring about a measure of gladness, or breathing space, to students, particularly to individuals who are scheduled to present their reports before the class. The feeling of glad-ness further swells when the students learn that there is a postponement of an examination, since this would give them more time to prepare adequately for it.

In cases when the professor would be absent, the student's major con-sideration would be the cost in going to school. A text message received before one left for school could save precious cash for transportation fare, not to mention the time and effort involved in the exercise.

In their diaries, the students declare that they are encouraged to study harder when their professors send them text messages. This inspires them to improve their class performance and, subsequently, get high grades.

A professor who sends text messages to students is characterized as "approachable," and therefore, students can easily consult with him/her regarding their lessons. The students would feel comfortable in the pres-ence of their professors, and not feel intimidated. As a result, the line of communication, for the purpose of learning, is established.

Having an open line of communication and direct contact with the professors makes the studying of lessons easier. The students feel free to ask questions, discuss with the professors and clarify vague concepts. Although these interactions are done face-to-face, the mobile phone fulfills its role of facilitating and making the discussions happen.

The Student-Professor Gap

The present crop of UP students was born at a time when massive technological changes were leading toward digitized activities, intercon-nectedness, etc. These students have grown up in a world of a global information economy, which has "resulted in a generation of learners who think and process information differently" from their parents and professors. Their world differs significantly from that of their profes-sors' (Benson 2008).

Benson (2008) argues that the professors have limited use of digital tools for their teaching activities. Their students, meanwhile, are more advanced, adventurous, creative and continually experimenting. They are curious and patient in learning new modes of doing things.

Although Benson (2008) says that there are early adopters and innova-tors among professors, most of them still prefer to continue with the old ways that they experienced when they were students themselves. Most of them still prefer face-to-face communication over the present practice of dispatching text messages through the Internet's Yahoo! Groups. As one of these professors would say, "Personal talks are always clearer."

Benson asks: "Is it the responsibility of the student to adapt to the educational tax or should the professor adjust his/her pedagogy to align with the world of the student?" Surely, they have to learn new ways of teaching.

The wide gap between professors and students became evident especially during the interviews as research informants gave divergent views. The professor-informants claimed that: a) the mobile phone offered very minimal help in improving their teaching style in terms of effectiveness;b) the mobile phone was not an essential part of their teaching—neither of their lessons, nor of the pedagogy;and c) profession-wise, the mobile phone was not as necessary to them as it was to the students.

Some reasons cited by professors for preferring other modes of communication, instead of the mobile phone, are as follows:

> I prefer other ways of contacting the students. In my view, sending text messages to them would be too informal for comfort. I think the e-mail would be more formal. The voice call, it's too expensive. It would also entail costs. At least, with the e-mail, you would know if there's a failure of delivery, right? Hence, the best for me would be the face-to-face communication.

However, the professor-informants acknowledge the ease that the mobile phone affords students and faculty members especially when there are urgent matters to discuss. In these cases, the mobile phone provides the initial phase from where further discussion would occur. The communication methods, featuring the face-to-face interaction, Yahoo! Groups and the landline phone, simply become superfluous.

Campbell (2002) posits that "age" is a function of a positive attitude towards the contribution of the mobile phone to learning. In his study, the youth have a more tolerant attitude towards using mobile phones in the classroom. He argues (2002: 14):

> This finding for age is consistent with previous research demonstrating that young people tend to have very positive perceptions of the technology and regard the mobile phone as an important tool for social connection.

Echoing Campbell's view, this study has found that student-informants think that the mobile phone contributes immensely to their academic activities. They cannot seem to go to class without their mobile phones. It is their link to their classmates and to information about class matters.

Why Professors Do Not Use Text Messaging

The mobile phone's limited capacity hampers the professor from elaborating on answers to a student's questions. Thus, the student is instructed to meet the professor in a face-to-face discussion. Meanwhile, the professor does not relish sending a long text message since he/she gets tired doing it. Neither does he/she want to be bothered with it.

The generation gap illustrates why professor-informants refuse to participate in text messaging. They are not fond of sending text messages to their students. In contrast, their students are frequently observed to be constantly holding their mobile phones and sending text messages. For the professor, the face-to-face interaction is better—especially in discussing topics, class policies, and concepts.

The seemingly obstinate attitude of professors against sending text messages to students is hinged on propriety, or the need to maintain "professional distance." They think, rightly or wrongly, that this kind of attitude would lead to, or command, respect from the students.

Cases of text messages of the romantic type have been cited, wherein professors become recipients of love notes through text messages from their students. Hence, to thwart such cases of students attempting to court their professors through the mobile phone, or inviting them to become text-mates, professors generally make it a policy not to divulge their mobile phone numbers to their students.

Other reasons cited are as follows: the mobile phone may be used to abuse professors, particularly when students dare to take advantage of their familiarity with professors in order to obtain a better grade; a professor has a life outside his/her world of teaching, and mobile phones to them are private spaces; giving one's number allows possible intrusions into their privacy as the giver becomes available anytime of the day, wherever he/she may be; mobile phones become irritants, especially when voice calls are made.

There are a few cases in the UP where youthful professors share their mobile phone numbers with their students. The number-sharing practice seems to emanate from their awareness that they are unable to spend sufficient time in the campus due to other commitments elsewhere. Hence, to compensate, they try to make themselves "available" through the wonderfully convenient mobile phone.

Unwritten Class Policies and Cultural Practices

The Filipinos' native and colonial tradition of respecting elders and legal or religious authorities is still very much a part of their modern lives, respect-bordering-on-reverence for professors, included. This is particularly obvious in the way Filipino students send text messages to their professors. In deference to their professors, the students carefully observe the following:

Only for Class-Related Matters

Students may send text messages for class-related concerns, but not for personal matters. Allowed are queries requiring only brief replies from their professors. Otherwise, their professors may direct them to use the Yahoo! Groups. A small number of students may advance to the text messaging stage, *i.e.*, on topics extending beyond class matters, especially when the students have already developed friendly relations with their professors.

This would normally happen after the semester ends and the students are no longer enrolled in classes under the concerned professors. According to the students, once this stage of openness between the professor and the student is reached, learning is facilitated. "There is blurring of authority lines," says a student.

Meanwhile, a professor looks at text messaging as opening up the Filipino concept of "*loob*" or inner self that radiates into a shared experience. It leads to unity and easy understanding, which then fosters an ambiance that is conducive to learning.

Be Formal: "No" to Text Lingo or Abbreviations

Although the mobile phone text lingo is generally regarded as an informal way of sending messages, the students negotiate its use and project formality by using complete sentences, or full text, in their text messaging with professors.

Professors prefer their students *not* using the "text language" and object to confusing spelling. The token use of vowels may reduce the message's length for a quick "send," but it somehow distorts the message and renders the substance ambiguous.

Some professors purposely avoid using the text language, lest they present themselves to their students as being sloppy or unprofessional, and not exactly deserving of professional courtesy.

This jibes with some students' belief that correctly spelled words and complete sentences are signs of respect. Thus, they should use "official-looking" text messages and apply correct and basic sentence construction.

I find these unwritten rules reinforcing the different ways that students send text messages: one for fellow classmates, and another for professors.

Professional Distance and Initiating Text Messaging Interaction

The messages between a student and a professor are usually short and simple, unlike the rather protracted and confusing text messages amongst students. In most cases, the student would initiate, and close, the sending of messages.

The rare times when professors would initiate the text messaging interaction are to return a reply to their students' queries, if necessary. There is no regularity in exchanging text messages with their students. This implies that, even in the mobile phone, there should be professional distance observed and maintained between the professor and the student.

Thus, propriety and professionalism could be the underlying reasons for the professor's terse and short types of messages. Everyone seems to expect that both students and professors should observe a professional (or respectful) distance.

But, for those who may have already developed friendly relations with their professors, the substance of their text exchanges changes and expands beyond mere academic concerns. For instance, students may send jokes, inspirational quotations, or quizzes, to their professors.

Text messages to a professor about one's inability to attend a class show due respect, and that the professor's class is not being taken for granted. Informing one's professor makes the student feel better, and he/she entertains the hope that the professor would allow him/her to catch up on a missed, or previous, lesson.

Text Forms: Brief and Respectful

Text messages do not go deep in terms of serious relations, and normally become an appointment medium for an extensive discussion of lessons. Professors expect due respect from students and they anticipate the words, "sir" or "ma'am," and "*po*" and "*opo* (Filipino words of respect usually addressed to elders) to be present in the messages.

Due to the sheer number of students, professors are wary about their Inbox becoming loaded with messages and about the costs for text messages to be sent to numerous students. They do not like costly and demanding calls and do not welcome a tone of urgency and pressure accompanying their incoming text messages.

Meanwhile, students think that it is discourteous for them to engage professors in text messaging. In case they have to, then they should observe the following: include a "Smiley emoticon" or "ü" to show a happy emotion; avoid ALL CAPS and exclamation points, among others.

Hence, considering all the foregoing, students should realistically expect a one-way communication, or even a no-reply, from professors.

Social Policy Implications

This study implicitly advocates that classroom teaching should benefit from the "exponentially increasing information explosion" (Benson 2008:1), particularly using the most ubiquitous form of media today, the mobile phone. Combined with another ubiquitous media, *viz.*, the TV, the mobile phone can deliver knowledge to students who may be marginalized by lack of resources. Since the mobile phone is in the hands of almost everyone, it then becomes a powerful tool in accessing knowledge and learning.

The Philippine government can enhance its educational system with an improved use and program for T2T with the mobile phone. The perennial problem of lack of classrooms, professors and limited access to education might be reduced or solved. Even when parents, due to poverty, are not able to send their children to school, their mobile phones can provide useful information to their children.

This implies the weight that should be given to enhancing the capability of the mobile phone in the future. Mobile phone companies invoking corporate social responsibility principles may well consider making available at cheaper cost, mobile phones that are Wi-Fi enabled.

Implications on Media Convergence

The possibility of the mobile phone's full utilization as an educational tool can materialize through the convergence of the mobile phone with other media, such as the Internet and TV. The T2T experience is succeeding because of the mobile phone's advanced features and larger capacity to receive, and be loaded with, video clips. It is, also, through convergence that knowledge is delivered in the classroom by plugging the mobile phone to a TV set and video lessons are played.

T2T could spell the difference between some learning and no-learning at all in a poor learning environment (Catangay 2009). It demonstrates how the mobile phone can be used effectively for learning and in bridging the educational divide, if mainstreamed in the educational system. The program shows how mobile phone can help uplift education in the Philippines.

The Mobile Phone Ban

To borrow from Katz (2005), the policies on banning the mobile phone from the classroom "did not emerge from thoughtful technology assessments, and mobile phones were not integrated into the school curriculum." In fact, the mobile phone created a predicament among professors and school administrators whether or not a strong policy should be enforced, since it disrupts the class and concentration or focus. But, if taken in a different light and the gadget is regarded as a learning partner, or likened to a book or laptop, then policies may be more lax and student-friendly.

Precluding the use of the mobile phone inside the classroom is tantamount to stifling the students' well-developed skills, which may be put to good use. Authorities, thus, may well differentiate the mobile phone use in social connections, games or play and academic setting. The "confused boundaries" between educational activities and games or social connections should be clearly defined.

The Professor–Student Gap

Retooling of faculty members to keep abreast with new technologies and changing their attitude towards texting students will narrow the gap between the professors and students. Adapting Benson's (2005) view, this would likely lead to the maximum utilization of the mobile phone as a learning partner and address the marginal impact of technological changes on teaching and classroom pedagogies.

Conclusion

The study argues that the mobile phone, as "notifier," facilitator, discourse venue, class interaction extension, etc., contributes to the students' learning. The mobile phone has thus assumed various roles in stimulating students to perform more efficiently in their academic and socio-cultural activities.

While the mobile phone with its varied uses exhibits its technological prowess and versatility, controversial issues inherent in its use adversely affect both students and professors. Nonetheless, students and professors negotiate the use of the mobile phone and cope with its technical limitations or imperfections by a) linking it with the Internet, b) maximizing its features, and c) utilizing it as a relay tool.

As a virtual enabler, the mobile phone has enhanced the potential of an open communication, or relationship, between professors and students, thus improving the learning ambiance and making academic and socio-cultural life more tolerable and convenient, if not fun.

Nonetheless, a significant finding of the study is what I regard as the digital gap between professors and students. Unlike the generally youthful, dynamic and creative students, the professors seldom make attempts to discover for themselves the mobile phone's many uses and benefits.

They are simply too busy and engrossed in their serious undertaking of formal, and perhaps traditional, teaching. Naturally, or inadvertently, they fall behind their students in making themselves more productive through a patient and creative use of the mobile phone.

As principal stakeholders in the learning process, the students and professors should infuse dynamism in their particular roles and functions to assure a balanced, as opposed to lop-sided, learning process. This would probably result to a mutually satisfactory and beneficial experience.

References

Alfonso, Herminia Corazon (2001). *Socially Shared Inquiry: A Self-Reflexive, Emancipatory Communication Approach to Social Re-search*, QC: Great Books Trading.

Benson, Spencer, Bradford Anna, Eubanks, David et al. 2008. *The Next Challenge and Frontier: Digital Learners and Digital Teachers*. University of Maryland Center for Teaching Excellence. Paper presented at the UNESCO Pre-Conference on World Education. Macau, China.

Braid, Florangel Rosario, Tuazon, Ramon and Gamolo, Nora. 2007. *A Reader on Information and Communication Technology Planning for Development*. San Juan, Metro Manila, Philippines: Asian Institute of Journalism and Communication (AIJC) and United Nations Educational, Scientific and Cultural Organization (UNESCO) National Commission of the Philippines.

Campbell, Scott. 2002. Perceptions of mobile phones in college classrooms: Ringing, cheating, and classroom policies. Department of Communication Studies, University of Michigan.

Catangay, Marisse. 2009. "Use of Mobile Phones for Quality Education" in *Philippine Daily Inquirer*. April 4, 2009.

Chipchase, Jan. 2008. "Reducing Illiteracy as a Barrier to Mobile Communication" in *Handbook of Mobile Communication Studies*. Edited by Katz, J. Cambridge: MIT Press.

Katz, James. E. 2005. Mobile phones in educational setting. In *A Sense of Place: The global and the local in mobile communication*. Vienna: Passagen Verlag.

Quirante, Aleli. 2006. "Texting, Etiquette and Social Relationships in Teen's Interpretive Communities" in Kim and Lee (eds) *Mobile and Popular Culture*. Paper delivered at the Asia Culture Forum, Gwangju, S. Korea.

Portus, Lourdes, Quirante, Aleli, and San Pascual, Maria Rosel. 2006. *ICT Literacy of Households and Individuals in Metro Manila*. An unpublished paper prepared for the Commission on Higher Education, Department of Education. Quezon City: College of Mass Communication, UP Diliman.

14

Lifeworld Keys and Intractable Objects: Privacy, Politics, and Mobile Symbolic Meanings in Italy

Matteo Tarantino

The social meaning attributed to communication interceptions has had considerable impact on the social shaping of telephonic technology, influencing all four levels identified by Campbell and Russo (2003): *"hardware (i.e., handsets), software, network systems, and service."* For example, with regard to landline phones, the perception of interception as a threat by concerned businessmen helped pushing towards the substitution of human operators with electromechanical switchboards (Mercer, 2006, p. 69). Similar perceptions were active in the switch from older mobile standards such as TACS (Total Access Communication System) and AMPS (Advanced Mobile Phone System), which could be intercepted using an amateur-level radio scanner, towards digital, encrypted standards such as GSM (Group Special Mobile / General System for Mobile Communication) (Van der Arend, 1988).

At the same time, the perception of interception as a valuable *tool for security* led governments to implement measures for enabling Lawful Interception (LI) of communication. Throughout the years, communication infrastructures have been redesigned to maintain LI feasible for law-enforcement operators. The most recent examples of these measures can be found in the CALEA standard in the United States and the ETSI standard in European countries. As a reaction, this prompted the design and commercialization of "interception-proof" handsets and software, thus influencing the "hardware" and "software" level.

To grasp the process of negotiation of the social meaning of interception is therefore important to properly understand a number of processes involved in the social shaping of the mobile phone technology.

The debate appears to revolve around two main axes: privacy/ security and public/private.

The privacy/security axis deals with the *collective* relationship with surveillance technology. Throughout the twentieth century, it has been the main frame of so-called "wiretap debate" (Zeni, 1949; Brown, 1960; Scott, 1968; Kleindienst, 1986; Lind & Otenyo, 2006; Yung, 1996; Diffie & Landau, 2008). The central issue of this debate is the trade-off between limiting privacy and ensuring security against threats to society, such as organized crime and terrorism. Research shows a predominant influence in this regard by the third-person effect (Davison, 1983; Perloff, 1993). In other words, interference into privacy for security purposes is generally accepted because it is conceptualized as *other people*'s privacy. Perry R. Morrison identified the psychosocial factor driving blind (i.e., careless of the social cost) recourse to electronic surveillance solutions in the attribution of the character of "*intractability*."

In fact, this is the essence of the technological embrace that seems to mesmerize our current thinking: When faced with an apparently intractable human problem, our habitual response is to seek the building of a technological system or artifact (Morrison, 1986).

According to the Random House Dictionary, "intractable" means "not easily controlled or directed; not docile or manageable; stubborn" but also "hard to shape" (as in a metal) and "hard to treat, relieve or cure." According to Morrison, when this character is attributed in social discourse to an object perceived as escaping or stalling the rules of a system (and therefore assumed as being impossible to *cure*) electronic surveillance measures, such as interception, are more likely pass as a viable for the social groups adhering to that discourse. Such is, in many countries, the case of the "terrorism" object. To understand properly social discourses about interceptions, it appears therefore worthwhile to examine what (and how) is conferred the attribute of "intractability" in a given context.

The second dichotomy is the one between public and private, and refers to the *subjective* relation with communication technology. Research has insisted much on the "personal" nature of the mobile phone artifact. Cross-culturally, the mobile is considered by users as an expression and extension of the body and of personal identity (Campbell, 2008; Katz and Aakhus, 2002; Katz and Sugiyama, 2006). This very "personal" nature of the mobile phone (reinforced by their holding material traces of us, in the form of call lists, SMSs and sometimes photos, videos and so on) makes them an ideal key to enter a subject's private life, to access, we might say, the subject's life-world. We can see traces of this in the fact that the mobile phone is held as a highly private artifact: people can react violently to unwanted access or manipulation of mobile phones (Kiesler, Gant, & Hinds, 1994).

Scope and Methodology

This chapter will discuss the negotiation of the social meaning of interception in Italy, by tracing the most salient features of the discursive space wherein this meaning is negotiated. The selected case study is the "Genchi Archive" scandal of 2007-2009, which will be analyzed in detail. Other interception-related cases could have been selected—for example, the 2006-2008 "Telecom Security" scandal, which revealed the security office of Italy's greatest phone operator to be a covert hub for illegal interceptions used for blackmail and profiling, part of wider illegal secret service operating in Italy (see Sidoti, 2008). However, the Genchi Archive case appeared to focus much more on the practice of interceptions than its counterpart, whose discourses in Italian media mostly focused on the profiling activities.

The analysis draws from a sample of newspaper articles. For quantitative statistical purposes, a total of 7588 wiretap-related articles have been gathered in the years between 1992 and 2009 from the two major Italian newspapers, *Il Corriere della Sera* (n = 3066) and *La Repubblica* (n = 4492). In-depth qualitative analysis has been performed on a subset of articles (n = 208) covering the Genchi episode. The sample also included articles from right-wing newspaper *Il Giornale* and left-wing *L'Unità*, periodicals *Micromega* and *L'Europeo*, blogs from relevant commentators (such as highly influential Beppe Grillo), and YouTube videos. All translations from Italian are the author's.

Quantitative Findings

Assuming the number of articles related to a topic as an indicator of its salience in the public agenda, the salience of interception appears to have considerably grown in the last years (see Figure 14.1). The total number of articles in the national press has grown from 242 in 1992 to 334 in 2000 to 617 in 2009, peaking with 945 in 2006. The patterns of the two newspapers appear similar before 1999 (see Figure 14.1), with the Corriere della Sera in the dominant position. After 1999 the overall number of interception-related news grows significantly in both periodicals, peaking in 2006.

Starting from 2000, *La Repubblica* started to outnumber *Il Corriere della Sera* in interception-related news. When we consider that the Berlusconi governments (2001-2005, 2005-2006 and 2007-present) have been criticizing the practice of interceptions, this variation appears coherent with the political orientation of the two periodicals. Along with fellow leftist periodical *L'Unità, La Repubblica* has always manifested a strong opposition to Berlusconi (Hibberd, 2007). It appears possible that both attempted to capitalize on the (potentially critical) aversion to interceptions of the Berlusconi government. This was not the case of *Il Corriere della Sera* (although not to be considered an explicit supporter of Berlusconi, which instead is *Il Giornale*) which has always followed a rather non-partisan, if conservative, line (Tungate, 2001, p. 111).

**Figure 14.1 Number of Interception-Related Articles in the
Two Major Italian Newspapers**

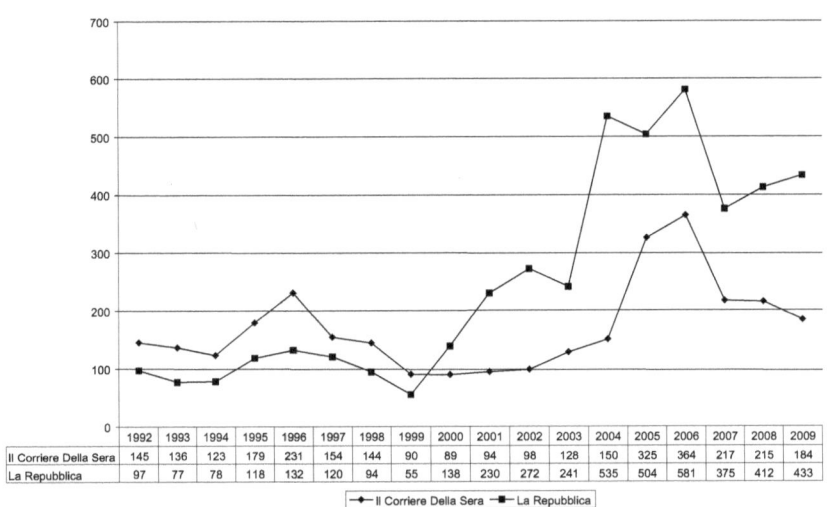

	1992	1993	1994	1995	1996	1997	1998	1999	2000	2001	2002	2003	2004	2005	2006	2007	2008	2009
Il Corriere Della Sera	145	136	123	179	231	154	144	90	89	94	98	128	150	325	364	217	215	184
La Repubblica	97	77	78	118	132	120	94	55	138	230	272	241	535	504	581	375	412	433

━◆━ Il Corriere Della Sera ━■━ La Repubblica

The proportion of Genchi-related news appears to follows a slightly dissimilar pattern. In *La Repubblica* Genchi appeared in 10.6 percent of wiretap-related news in 2007, 4.3 percent in 2008 and 14.3 percent in 2009 (n = 40, 18 and 62 respectively). The *Corriere della Sera* percentages are 7.3 percent in 2007, 3.7 percent in 2008 and 34.7 percent in 2009 (n = 16, 8 and 64 respectively). This means that while *La Repubblica* refers to interceptions much more than *Il Corriere della Sera*, when the debate on the scandal raged on in 2009 Genchi was mentioned more on *Il Corriere della Sera* than in its leftist counterpart.

Background

First regulated by law in the 1913 procedural code, phone interceptions were systematized under Mussolini through the Rocco Laws (Guspini, 1973). After the war, the Christian Democratic Party, the center-right formation who dominated the Italian political scene from 1945 to 1994, enacted special measures (especially Law 98 of 1974 and law 191 of 1978) against the threat of political terrorism during the so-called "lead years" (1964-1984). These measures allowed prosecutors a great deal of liberty in enacting surveillance against suspected associated crime and terrorist organizations. The use of interceptions was again largely publicized in the 1992-1996 period, known as "Tangentopoli," from the codename of an investigation on public corruption that revolutionized Italian postwar politics by destroying the Christian Democratic and Socialist parties. After the 2001 terrorist attacks in the United States, Italy passed new security reform lowering interceptions' requirements to a mere prosecutor's warrant (Regoli & Hewitt, 2009, p. 175). As a result, electronic surveillance started being used in a large number of investigations, and the number of interceptions skyrocketed, almost doubling yearly since 2001 (ibid.). Moreover, interception transcripts have been published on newspapers without much restraint or regulation. All of this, combined with the new ubiquity of phone communication due to mobile phone penetration, created what Sidoti (2008) called a regime of very high "visibility of illegality," which has arguably no peers in the world. Interestingly, this "new visibility" is almost entirely the product of voice communication.

In 2002, the Berlusconi government started to express strong reservations on interceptions (Galluzzo, 2002). In 2003, a report by the Max Planck Institute ranked Italy first among Western nations with 76 intercepts every 100,000 citizens (Albrecht & Kilchling, 2003). In February 2005, the report was used as a political tool by the second Berlusconi government (2001-2007) as then-Justice Minister Roberto Castelli deemed the number of interceptions as "too high" and "too costly." From 2005 to 2009, the Berlusconi governments attempted to pass several bills restricting the recourse of interceptions. The latest iteration of such measures, called the Alfano Reform, limits interceptions to a handful of crimes, curtails their duration to a maximum of 60 days, complicates their procedural requirements (only "evident proofs of crime" justify the measure), is presently (as of February 2010) being examined by the Italian Senate (*Norme in materia di intercettazioni telefoniche…*, 2009).

A Case Study: The Genchi Archive

The Genchi Archive controversy first appears on Italian media in 2007: Its intensity dips in 2008, and picks up again in 2009/2010. Its main protagonist is Gioacchino Genchi, a police consultant based in Palermo with an expertise in communication interceptions—precisely, the analysis of phone call records. Even before the scandal, Genchi's long career (first as a member of State Police and subsequently as a consultant) had made him a recurrent, if relatively unknown, presence in the columns of Italian newspapers since the early 1990s, mostly commenting interception-related investigations (see, for example, Cavallaro, 1992).

In July 2007, Clemente Mastella, justice minister of the second Prodi government (2006-2007) is involved in an investigation on a hidden cartel of entrepreneurs and politicians. His involvement is allegedly proven by the analyses of phone records performed by Genchi, then a consultant of the Potenza prosecutor (Bianconi, 2007). Mastella reacts vehemently and orders to investigate both Genchi and the prosecutor. In January 2009, under the new Berlusconi government, Genchi is accused of owning an illegal archive with the raw data of all of his investigations, gathering data intercepted from thousands of citizens (Genchi's archive is seized in February 2009 and returned to the owner in April 2010, with large sections being classified).

Berlusconi himself, although not directly involved in the investigations, speaks of the Genchi Archive as "possibly the greatest scandal in the history of the Republic" (Casadio, 2009). The case is used by Berlusconi himself and by its government to support the necessity of a restrictive reform of interceptions; Berlusconi states that

> we must be determined and avoid that the recourse to this system, deemed as exceptional by out Constitution, continues. (...) We must set definite boundaries, safe for citizens (...) Interceptions must be allowed only in presence of evident proofs of crime, and limited within a thirty days window, with a possible fifteen days extension. We will not renounce to our privacy anymore (Casadio, 2009).

An intense debate subsequently ensues in Italian media, already sensitive to the topic because of the discussion on the proposed Alfano reform bill.

A first evident feature of this debate is a process of explicit *heroicization* of Gioacchino Genchi enacted by pro-interception voices. This process can be observed in *La Repubblica*, and other periodicals of the same orientation (such as monthly magazine *Micromega*). For example, in describing Genchi's headquarters, *La Repubblica* underscores its being a place "other" with respect to their immediate surroundings (Sicily) and, by extension, to all of Italy:

> Who really is the super-consultant from Palermo who angered the Minister of Justice?
>
> To answer this question we must come to Sicily (...). Outside is the eternally disheveled traffic of Palermo and a silent seller of bread, fritters and croquettes. There is incessant noise of car horns, voices, engines. Inside, the silence of a Swiss clinic. We must descend to the basement of the building to discover the aseptic air of Genchi's offices, five hundred armored square meters which make you feel like being in Locarno, or Geneva. Monitors everywhere. Servers. Telephones. Files. Green plants unhappy to be underground, under neon lights (Barbacetto, 2007).

The man operating the premises, Genchi, is described with the features of an alchemist or a wizard, transmuting worthless elements into precious ones:

> He, Genchi, speaks little and listens a lot. Especially phone calls, especially in headphones. But his specialty is phone records, computerized lists of calls. He throws them into his computer, analyzes them, shakes and squeezes them and eventually pulls out relationships, recourses, complicities (Barbacetto, 2007).

The power of interceptions is never downplayed by the pro-Genchi discourses. Quite on the contrary, this power is often exalted. However, pro-interception voices to present this *magic* as safe because of being wielded by the State itself through the heroicized Genchi, the "good interceptor" (Montelli, 2007). Genchi (who is, for all intents and purposes, a private consultant) is continuously described as a loyal "servant of the State":

> Can anyone lie so shamelessly and use false allegations to try to destroy the reputation of a man like Gioacchino Genchi, who has always served and continues to serve the State? (...) Gioacchino Genchi must be defended at all cost by everyone who cares for the future of Justice in Italy (Borsellino, 2009).

In this mythic representation of interception ad of interception operators, the source of the "magic" of interceptions is precisely the lifeworld-penetrating power of the mobile phone. The potential violation of the "intimate space" of mobile phone, instead than inducing anxiety, reaffirms security, as the privileged access granted by mobiles to individual life-worlds, together with the self-evident ubiquity of the device, establish the conditions for a *truly capillary justice*.

This representation is capitalized by pro-interception discourses: the "tracking" and "shadowing" capabilities of phone interceptions (traditionally capitalized by anti-interception discourses), are covered

extensively by these voices, exalting the new "regime of visibility."
Consider the following passage, quoting directly Genchi:

"The three-dimensional cartographic visualization of points of interest and of
coverage areas of GSM cells enabled the tracking of movements and migra-
tions of the mobile handsets belonging to investigated subjects who performed
telephonic connections." It is as if Genchi performed "reverse-shadowing,"
retracing the steps of the investigated subjects or their interlocutors through the
tracks let by mobile phones. This mass of data was then presented to the prosecutor "through the realization of a Web system protected by a cutting-edge
encryption and access management system," which even includes "valorization
of GSM cells' areas of coverage through the cartography of places, performed
through a complex interaction among web applications Google Maps and Google
Earth" (Barbacetto, 2007).

Along with the mythopoesis goes a process of definition of the "intractable objects" justifying and motivating the "magic" of interceptions.
We have here an evident forking between elite (enacted by visible and
recognizable commentators) and grassroots (i.e., emerging in spaces
such as Internet discussion forums) discourses. In elite discourses, while
organized crime, terrorism and corruption are often mentioned, the focus
appears to be on *sexual predators*. The following excerpt is taken from
the radical leftist newspaper *l'Unità*:

The brilliant operation [the arrest of a gang of rapists] (…) was made possible by
cross-referencing wiretaps and phone records: what Gioacchino Genchi, public
servant forced to defend himself from despicable political attacks, has been doing
for 20 years. But with the new law on interceptions, who will intercept a bunch of
alleged rapists will have to hope that they betray themselves at once: after the first
45 days, we must stop listening. The new "genial" bipartisan idea allows interception for a month and a half, and no more. Either one is caught saying "let's rape her"
or naming the victim within 45 days, or no one will ever know who he is. And the
rapist will commit his crime in all due peace. After the State massacres, we have
State rapes (Travaglio, 2008).

A post on the most-read blog in Italy, operated by caustic satirist Beppe
Grillo, opens with the following statement: "Wiretapping *helps us to
capture delinquents. Less wiretapping equals more rapists, pedophiles
and* corruptors walking around freely" (Grillo 2009). In another example,
Giuseppe Cascini, secretary of the National Magistrate Association (and,
as such, spokesperson of a large quota of the Italian judiciary) writes an
open letter (published on the front page of *La Repubblica* on January 12,
2009) criticizing the proposed bill by depicting a possible future scenario
where, under the new bill, sexual predators have a much higher chance
of avoiding capture. The title is "How a pedophile will escape justice"
(Cascini, 2009). Even in the Parliament, pro-interception voices put
sexual crimes on the forefront: consider the following excerpt from the

official speech of opposition member Federico Palomba, commenting the interception reform:

> How will it be possible to repress the cases of sexual violence, corruption (…), rob-bery, extortion, even homicides, with many, too many obstacles on the way of such operations? Crime will be thankful (Palomba, 2009).

Other examples abound, as a variety of social actors took similar positions (see, for example the voice of prosecutors expressed in Stefa-nini, 2008 and Lavinia, 2009). Grassroots discourses appear instead to privilege politics as the "intractable object," and seldom mention sexual predators. For example, the 2,077 comments (as of February 2, 2010) to the YouTube video containing Genchi's interview are largely dominated by pro-Genchi voices who focus solely on politics. Often, the comments are addressed directly to Genchi, calling for its intervention on what are described as Italy's rotten politics. The tenor of the comments is the following: "*Genchi is an hero, and history eventually will prove that what he does is right*" and "*Please kick them all in jail, they are all just a bunch of shameless mafiosi and ruffians.*"

After an initial low-profile phase, Genchi eventually starts to present himself in the same fervent (for some, populist) light, championing the "blindness" of justice allegedly granted by technology. In one interview he states that

> Law is equal for everyone. We are all to submit to the rule of law! It has to be clear; they (politicians) must understand this. When someone does so much as get any close to those gentlemen, they rebel and destroy anyone who simply dares to do his job. Italian people understood this (Genchi, 2009).

Anti-interception discourses appear much more traditional in structure and content than their counterparts. *Il Corriere della Sera* appeared to assume an alarmed and critical position towards Genchi. Consider the following excerpt from an editorial from Pierluigi Battista, one of *Il Corriere della Sera*'s highest-profile commentators:

> Do you feel safe in a country where a super computer expert, an attorney's consultant with the power to intercept telephones, happens to own more than 500,000 records about illustrious and less illustrious Italian citizen? Aren't you a little worried—even before you know what the hell [sic] may be written, recorded and noted in those disturbing, mysterious records—to know that the "information bounty" of Mr. Gioacchino Genchi includes (…) information about "non-investigated" people? (Battista, 2009).

Traditionally, anti-interception discourses reference totalitarianism, often drawing upon the imagery of twentieth century's totalitarian re-gimes. The case applies here as well, as the article draws a parallel with the German film *Das Leben der Anderen* (Henckel von Donnersmarck,

2006), which tells the story of a zealous and ruthless East German STASI (secret police) officer during the 1960s:

> How many Italians, largely "non-investigated persons," have seen a film like *The Lives of Others*, a realistic representation of the astounding arrogance implied in unchecked state espionage? Do we really believe that democratic Italy is by nature and vocation immune to the threat of such a virus? (Battista, 2009).

However, the images of giant, impersonal apparatuses already well established in the collective imaginary are described as being even more dangerous than in the past because of being *out of control*. This would be due to the chaotic state of Italian politics, which allows unrestrained leaking of information on the media:

> While in Germany the proceedings of the Great Communist Ear remained confined in the secret rooms of a despotic political police, in Italy the records of "non-investigated persons" can constantly end on the pages of newspapers? (Battista, 2009).

The central idea is that *anyone* (including the reader) can be caught into the mayhem, especially due to the involvement of the media system:

> Can't we understand that the point is precisely the havoc that can be unleashed on the privacy of "non-investigated" people no longer shielded from intrusions, fabrications, gossip, hostile revelations, threats to reputation, and violations of the fundamental right to privacy? (Battista, 2009).

In this light, the "magic of interception" becomes evil, and while the alchemical imagery remains it now produces injustice:

> [T]he "Genchi System" [is] the shaker of phone records, names, contacts, in which in the end everyone knows everyone else and (therefore) anyone can be accused of anything. Thousands of name, millions of contacts (Fazzo, 2009).

A different kind anti-interception discourse (enacted only by a minority of voices) attempts instead to deconstruct the heroicized image of Genchi:

> Genchi has been sanctified in the name of interceptions, precisely what he denies having performed. This is the paradox: while Genchi is indicated as Italy's superspy, hiding in his house in Palermo the secrets of the country, he denies it and says to have followed orders coming from magistrates, and at any rate of never having worn headphones or pushed the REC button (Filippi, 2009).

However, the power of pro-Genchi elite discourse in defining the intractable object as sexual predators is such that competing discourses are forced to defend themselves on the same ground: for example, in an article in which Domenico Vulpiani, commander of Italian Postal Police (which deals ICT-related crimes), states that, "even without Genchi," Italian Police arrested its "fair share" of pedophiles (Mancini 2009).

Shaping a New Market

The SST paradigm predicts that the social meanings of a mobile-related practice influences the hardware and software levels of the mobile phone technology. Such appears to be the case of Italy. Consumer interception and anti-interception systems can be found on Internet specialized shops. Products range from Symbian-powered smart-phones loaded with software for remote environmental listening, to bug-detector hardware and, what is more interesting for us, to "anti-interception phones."

The most famous "crypto-mobile" in Italy was developed by a Turin company called CasperTech. In 2004, it introduced the "crypto-phone," a GSM-based "secure" system called HTC P3450 (later reissued as two separate products, Easy Cryptech and Cryptech, featuring different degrees of encryption). In 2007, Caspertech established a partnership with the Association of Italian Congressmen to supply its anti-interception phones to politicians at reduced prices. In the press release accompanying the event, the company identified its mission as "satisfying the needs of the military and political world after numerous interceptions have been published on newspapers." The price of Caspertech's products is around 1,000 euros per handset. Reportedly, the company sold 10,000 as of August 2008 (Stancanelli, 2008).

Another Italian company operating in the sector is EndoAcustica. Endoacustica produces and sells online (through its so-called Spystore) three different solutions: The Safe-And-Talk Symbian encryption software (compatible with Symbian-powered handsets), a GSM-CRYPTO terminal (which performs stronger encryption than the S&T software) and, finally, its top product, a "Stealth Phone" which randomly generates a new IMEI code for each call, thus allegedly making interception "impossible." When in 2009 Endoacsutica tried to enter the U.S. Market with Safe-and-Talk, the selling strategy was played on an "Americanized" version of the key-to-lifeworld attribution:

> Living with the fear that what we say might be heard by others, and used to harm us in business or in the family, has in recent years become a serious threat to our well-being, so serious that under the Bush administration, the US Senate passed a bill that guarantees retroactive immunity for those US phone operators which participated in the so called DSP, or Domestic Spying Program.[1]

The EndoAcustica homepage happens to exhibit an apparently inexplicable "anti-pedophile" banner which links to a dedicated subsection advising parents against the risks of the Internet as a dangerous place for unsupervised minors:

> **Protect your kids! Say stop to those who produce and commercialize pedo-pornographic material.** The Internet is an open window on the world, an ocean of infinite possibilities for our children that have the opportunity to grow and learn, sharing dreams and hopes for the future, at pace unthinkable until few years ago. **All of us adults have the responsibility for their curiosity, for their interest in the world. Let's educate them to distinguish. Don't let them be alone.**[2]

The content of this page appears completely unrelated to the rest of the website, but its function is clear once we place Endoacustica as a social actor inside the framework described above. EndoAcustica appears then attempting to shield himself from a powerful argument, which we may summarize as "if wiretapping is useful for apprehending pedophiles, then anti-wiretapping gadgetry is useful to pedophiles themselves." Almost certainly unwillingly, the strategy of the company appears to confirm the role of sexual predators as the "designated threat" to be fought by interceptions, thus eventually strengthening the pro-wiretap position.

Conclusions

In Italy, the negotiation of the social meaning of telephone interceptions is the product of the negotiation between actors belonging to three spheres: civil society, politics, and the judiciary. Interception-related scandals appear to re-enact this negotiation, which otherwise tends to slip into a state of silent acceptance. When the negotiation is on, a plurality of actors (pursuing various goals) enact complex discursive strategies depicting competing scenarios. The case study enables showed that both anti-interception and pro-interception discourses share a common root image: that of the mobile as a key to access individual life-worlds. While the privacy/security axis shapes much of the argument, it is the public/private axis that provides the backbone of the discourse about interception in Italy.

Anti-interception discourses evoke directly the intimacy with mobile phones (for example, commenting the Genchi case Berlusconi declared that "it is unacceptable not to be free to speak on the phone"). The thesis here is that since individual life-worlds always contain potentially-ambiguous elements (namely, words spoken and contacts reached) exposed to interception, and since the judiciary is depicted as a blind, technocratic machine, *everyone* (including the reader) runs the risk of indiscriminate repression.

Pro-interception discourses also evoke the intimacy of mobile phone, but appear to use it to depict a potential *ubiquity of justice* (with hints of social equalization, especially in grassroots discourses) and shift the anxiety on the intractable object itself.

This discursive field appears reversed with respect to the traditional configuration (as described, for example, by Posner, 2008): in Italy, leftist parties (and even more so populist leftist parties) defend interceptions, while right-wing parties appear to oppose them—all of this while maintaining agendas (and alliances) otherwise akin to those of countries such as the United States. (This configuration could be interpreted as a radicalization of Fletcher's thesis that the same elites who are more supportive of civil rights are also more supportive of electronic surveillance. The main difference lies in the fact that Fletcher's subjects held these opinions in private, whereas here we have a public, and even political position (Fletcher, 1989).)

It may be hypothesized that the reason for this awkward configuration is the presence at the very core of Italian politics of an intractable object: Silvio Berlusconi itself. The massive concentration of power in his hands, together with an intense cult of personality, creates an aura of "pure intractability" around his persona. In the opposition imaginary (but also, to some extent, in the collective one), he is described as someone who, simply, *cannot be dealt with* in any manner, a pure *anomaly* concentrating the legislative, executive, financial, and media power. Therefore, not only wiretapping would be justifiable against him; not only his immense amount of power would justify the annihilation of his personal privacy; but all of this is presented as a *necessary* measure, a form of control over an otherwise uncontrollable object, through the access to Silvio Berlusconi's life-world through its (or its collaborators') mobile phones. In this logic, the underlying, and often explicitly stated (especially, but not only,[3] by grassroots discourses) assumption is that the true rationale of anti-interception measures is to protect Berlusconi himself—not so much from specific allegations, but from this "last line" of public control. In this logic, not only are interceptions useful for treating the intractable, but also *their publication is a necessary part of the process*. We might say that these voices call on the mobile phone to be the cornerstone of a new regime of "visibility of power" fulfilling Thompson's (1995) promise of the "reversed Panopticon" where media enable a closer and tighter public control of political power.

Notes

1 See http://www.articlesbase.com/cell-phones-articles/a-cell-phone-encryption-software-and-you-can-talk-freely-930233.html. It must be noted that such tones are nowhere to be found on the company website itself.
2 http://www.endoacustica.com/protect_your_kids_en.htm.
3 Consider, for example, the following excerpt from an article titled "*To protect Berlusconi from interceptions: The press silenced, mafiosi favored*" published on the Marxist-Leninist party's website: "With the excuse of "granting the citizens' right to privacy and security", Berlusconi wants to close the game once for all with magistrates and journalists, by taking away from the former the important investigative tool of interceptions, and by closing the mouth of the latter" (*Per proteggere Berlusconi dalle intercettazioni telefoniche Imbavagliata la stampa, favoriti i mafiosi*, 2009).

References

Barbacetto, Gianni. (2007, November 2). Genchi, l'uomo dei telefoni, *Il Venerdì di Repubblica*.

Battista Pierluigi. (2009, January 26). Il nuovo spionaggio e le vite di tutti, *Il Corriere della Sera*, p. 24.

Bianconi Giovanni. (2007, October 20). Clemente, l' imprenditore e quel cellulare della Camera, *Il Corriere della Sera*, pp .8-9.

Brown, Peter Megaree. (1960). Great Wiretapping Debate and the Crisis in Law Enforcement, The. *New York Law School Law Review, 6*, 265.

Campbell, S. (2008). Mobile Technology and the Body: Apparatgeist, Fashion, and Function, in J. Katz (Ed.). *Handbook of Mobile Communication Studies*. Cambridge: MIT Press.

Campbell, Scott W. (2007). Perceptions of mobile phone use in public settings: A cross-cultural comparison. *International Journal of Communication, 1*(1).

Casadio, Giovanna. (2009, January 25). Berlusconi: in arrivo uno scandalo enorme, *La Repubblica*, p. 9.

Cavallaro Felice. (1992, November 28). L' allarme strage perso negli uffici, *La Repubblica*, p. 18.

Davison, Philip. (1983). The third-person effect in communication. *Public Opinion Quarterly, 47*(1). 1-15.

Diffie, William and Susan Landau. (2008). *Privacy on the line: the politics of wiretapping and encryption*. Cambridge: MIT Press.

Donohue, Laura. (2006). Anglo-American Privacy and Surveillance *The Journal of Criminal Law and Criminology, 96*, 3, 1059-1208.

Fazzo, Luca. (2009, January 27). Svarioni, amnesie e avvertimenti Lo show di Genchi, uomo dei misteri. *Il Giornale*.

Filippi, Stefano. (2009, February 6). Ora Santoro, Grillo e Travaglio "santificano" Genchi in tv. *Il Giornale*.

Fletcher, Joseph F. (1989). Mass and Elite Attitudes About Wiretapping in Canada: Implications for Democratic Theory and Politics. *Public Opinion Quarterly, 53*(2). 225-245.

Galluzzo, Marco. (2002, July 18). Intercettazioni e processi, l'offensiva è solo rinviata. *Il Corriere della Sera*, p. 8.

Galluzzo, Marco. (2009, January 25). «In migliaia sotto controllo Presto un grande scandalo." *Il Corriere della Sera*.

Genchi, Gioacchino. (2009). Le interviste del blog beppegrillo.it: Gioacchino Genchi, retrieved from www.youtube.com/watch?v=LbkUw0Bopmw.

Grillo, Beppe. (2009). *Caselli, Wiretapping and Penelope*, retrieved online at www.beppegrillo.it/en/2009/06/caselli_wiretapping_and_penelo.html.

Guspini, Ugo. (1973). *L'orecchio del regime*. Milan: Mursia.

Henckel von Donnersmarck, Florian. *Das Leben der Anderen*. Buena Vista International, 2006.

Hibberd, Matthew. (2007). *The media in Italy: press, cinema and broadcasting from unification to digital*. Open University Press.

Kasper, Debbie. (2005). The evolution (or devolution) of privacy, *Sociological Forum*, 20, 1, pp. 69-92.

Katz, James E. (2003). *Connections: Social and cultural studies of the telephone in American life*. Transaction Publishers.

Katz, James E. (ed.) (2008). *Handbook of mobile communication studies*. Cambridge: MIT Press.

Katz, James E. and Satomi Sugiyama. (2006). Mobile phones as fashion statements: evidence from student surveys in the US and Japan. *New Media & Society*, 8 (2). 321.

Kiesler, Sara, D. Gant, and Pamela J. Hinds. (1994). *The allure of wireless: Preliminary report on a trial of PCS telephony*. Carnegie Mellon University.

Lavinia, Gianvito. (2009, June 12), «Abbiamo arrestato ottanta pedofili Ora sarà impossibile», *Corriere della Sera*, p. 11.

Lind, Nancy S. & Otenyo, Eric Edwin. (2006). Administrative Agencies in a Technological Era: Are Eavesdropping and Wiretapping Now Acceptable Without Probable Cause? *International Journal of Public Administration*, 29 (14). 1397-1409.

Mancini, Lionello. (2009, January 30). Vulpiani: «Genchi? Meglio utilizzare le forze dell'ordine», *Il Sole-24 Ore*, retrieved online at www.ilsole24ore.com/art/Sole OnLine4/Italia/2009/01/criminalita-telematica-intervista-vulpiani.shtml.

Martirano, Dino. (2007, July 28). Cellulare di Mastella online. Lui: farabutti, *Il Corriere della Sera*, p. 5.

Montelli, Edoardo. (2007). Gioacchino Genchi Il Buon Intercettatore, *L'Europeo*, 2, pp. 21-24.

Morrison, Perry. (1986). Limits to Technocratic Consciousness: Information Technology and Terrorism as Example. *Science, Technology, and Human Values*, 4-16.

Norme in materia di intercettazioni telefoniche, telematiche e ambientali. Modifica della disciplina in materia di astensione del giudice e degli atti di indagine. Integrazione della disciplina sulla responsabilità amministrativa delle persone giuridiche, Italian House of Parties, Berlusconi-IV Government, presented on June 11 (2009).

Palomba Federico. (2009). *RELAZIONE DELLA II COMMISSIONE PERMANENTE (GIUSTIZIA) presentata alla Presidenza il 20 febbraio 2009 SUL DISEGNO DI LEGGE PRESENTATO DAL MINISTRO DELLA GIUSTIZIA (ALFANO)*. Italian parliament official document. Retrieved from www.camera.it/_dati/leg16/lavori/stampati/pdf/16PDL0020242.pdf.

Perloff, Richard M. (1993). Third-person effect research 1983-1992: A review and synthesis. *International Journal of Public Opinion Research*, 5(2). 167-184.

Posner, Richard A. (2008), Privacy, Surveillance, and Law, *The University of Chicago Law Review*, Vol. 75, No. 1. pp. 245-260.

Regan, Priscilla M. (1995). *Legislating privacy: Technology, social values, and public policy*. University of North Carolina Press.

Regoli, Robert M., Hewitt, John D. (2009). *Exploring Criminal Justice: The Essentials*. Jones & Bartlett Pub.

Sarzanini, Francesco. (2009, February 4). Genchi, nell' archivio un italiano su dieci. *Corriere della Sera*.

Scott, Hugh. (1968). Wiretapping and Organized Crime. *Howard Law Journal*, 14, 1.

Sidoti, Francesco. (2009). The Italian Intelligence Services in Thomas Jäger, Anna Daun (edited by). *Geheimdienste in Europa: Transformation, Kooperation und Kontrolle*, VS Verlag, pp. 78-99.

Stancanelli, Bianca. (2008, August 08). Intercettazioni Istruzioni per L'uso, *Panorama.*

Stefanini, Marco. (2008, July 17). "Intercettazioni fondamentali contro la pedopornografia", *La Repubblica.*

Stupro di Guidonia, presi quattro romeni Uno confessa. Folla infuriata: «Bastardi." (2009, January 27). *Corriere della Sera.*

Per proteggere Berlusconi dalle intercettazioni telefoniche Imbavagliata la stampa, favoriti i mafiosi, retrieved online at www.pmli.it/imbavagliatastampafavoritimafiosi.html.

Thompson, John. (1995). *The Media and Modernity: A Social Theory of the Media,* Stanford University Press.

Travaglio Marco. (2009, January 29). Zorro, *L'Unità.*

Tungate, Mark. (2004). *Media Monoliths: How Great Media Brands Thrive and Survive.* London: Kugan Page.

Van der Arend, Peter. (2008). Security Aspects and the Implementation in the GSM System, *Proceedings of the Digital Cellular Radio Conference*, Hagen, Westphalia, Germany.

Yung, Andrew W. (1996). Regulating the Genie: Effective Wiretaps in the Information Age. *Penn State Law Review, 101,* 95.

Zeni J, Ferdinand J. (1949). Wiretapping—The Right of Privacy versus the Public Interest. *Journal of Criminal Law and Criminology, 40,* 476.

15

Mobile Political Campaigns: The Nexus of Mass Content and Private Consumption

Chih-Hui Lai

Mobile phones have long been accorded importance as a personal communication medium because people are driven to satisfy their various communication needs on a constant basis (Katz & Aakhus, 2003). The widespread use of the mobile medium also makes it relatively easy to reach individuals with public messages, and thus gives rise to an emerging contested terrain in political participation, to the benefit of political campaigns and the public. In the widely discussed 2008 U.S. presidential election, mobile phones were used to an unprecedented degree (at least in the US context) as a novel element in the campaigns, most notably in support of presidential candidate Barack Obama. This was especially the case in facilitating political communication both among voters and between voters and candidates (e.g., Vargas, 2007; Katz 2009). During the 2008 election cycle 11 percent of text messaging (SMS) users reported receiving messages directly from a candidate or political party and more than one-third (36 percent) used text messaging to communicate with their social contacts about the campaigns, according to a Pew national survey (Smith, 2009). (In Pew's 2009 report, a comparison was made in terms of online media use by Americans for political communication and information seeking during 2004 and 2008 presidential elections. But there was no mention of text messaging use for these purposes earlier than 2008.)

Not only in the U.S., but also across the globe, the use of mobile phones for electioneering is growing. Examples include election advertisements through SMS in India and Hong Kong (Adelman, 2004), instant monitoring of voting irregularities by individuals with mobiles in African countries (Steere, 2008), vote reminders in Thailand's 2005 elections, the United Kingdom's 2001 elections and Italy's 2004 elections (Hermanns, 2008;

Mobiledia.com, 2005; Ward & Gibson, 2003), mobile versions of the candidates' websites in Australia's 2007 elections (Murphy, 2007), party candidate nomination through mobile voting in Korea's 2007 presidential election (Kim, 2007), and voter mobilization in Ukraine's 2004 Orange Revolution (Goldstein, 2007), to name just a few.

Due to some highly publicized examples claiming that mobile phones usage was pivotal in political campaigns, many thought the mobile would of itself lead to broader public participation in political contests, with the result that democratic processes would be strengthened and reinvigorated. Indeed, incorporating text messaging in campaigns has been linked to increased voter motivation and effective voter mobilization in several recent elections (Dale & Strauss, 2007; Han, 2007; Suarez, 2006). On the other hand, there is research that casts doubt on this premise, claiming that the impact of mobiles on political campaigns has been overestimated (Coleman, 2001; Lee & Willnat, 2006; Prete, 2007). Reflecting on these debates, this chapter aims to survey existing research to answer four questions:

1. What is the role of mobiles in campaigns,
2. Who is in control of mobile campaigns,
3. How does the influence of mobile campaigns take form, and
4. What are the outcomes of mobile campaigns?

Rather than providing an exhaustive review, these questions are delineated with the goal of building a practical lens to stimulate further discussion and research on the impact of mobiles in political campaigns.

Questions Asked about Mobile Campaigns

What Is the Role of Mobiles in Campaigns?

There are expectations that the use of information and communication technologies (ICTs) such as email, blogging, social networking sites and mobile phones can better motivate public participation and add new meaning to political campaigns (e.g., Anstead & Chadwick, 2009; Kreiss, 2009). Features of immediacy and mobility unique to the mobile phone make it an ideal medium for quick and simple organizing such as updating voters on the real-time campaign activity via their mobiles. The reliability of the messages, as perceived by mobile phone users, also helps infuse mass-produced campaign messages with personalized appeal (Sükösd & Dányi, 2003). But having seen these values attributed to mobile phone use leads us to ask: If there are useful and usable features inherent in mobile communication, why are the observations thus far inconsistent as to the impact of mobiles in political campaigns?

Societal, Political, and Historical Contexts Matter. In fact, despite many features of the mobile, caution must be used when interpreting the events of successful campaign mobilization made possible by SMS. These episodes may have been fueled by a historical background involving people who desired a radical change in the social and political environment (Coleman, 2001), as with the cases of Iran's 2009 presidential election and Ukraine's 2004 presidential election, which outweighed the primary effects of mobile phone use. Moreover, the impact of mobile phones in campaigns varies among societies featuring different structures of technological and political development. In Hong Kong, for example, where despite a few memorable instances of citizen-initiated voter mobilization and demonstrations occurred with the aid of advanced SMS networks and email, people generally are more likely to be apathetic relative to their politicians. These mobile-enabled political actions may be rendered marginal given the ingrained political system biases (incomplete democracy) (Yeung, 2007). Singapore, while having similar technical advantages of widespread ICT use yet a different type of submissive political culture, has not seen significant employment of these new technologies for political purposes (Lee & Willnat, 2006).

Mobile Campaigns within the Current Media Landscape. As one of the important sources for receiving political information, mobile phones are often seen as embedded in and interdependent with other media use such as television, newspapers, radio, and the Internet when it comes to public participation (Hermanns, 2008). During Spain's 2004 election, the successful voter mobilization effort was mainly ascribed to the formation of emergent networks of young voters using SMS, but the traditional media's exposure of the Spanish government's misrepresentation of the investigation of terrorist attacks also played a critical role (Suarez, 2006). SMS, along with other traditional and Internet technologies, also contributed to the successful debut of the fledgling Lok Paritran Party in India's 2006 election (Rajeev Gowda, Narayan, & Ollapally, 2006). In Taiwan, multimedia mobile technologies were gradually and seamlessly integrated as part of the converging media platforms for campaign communication in the 2008 presidential election. (This campaign activity was organized by a group of television broadcasters, newspapers and cell phone operators. The debate took place on February 24, 2008. The public was encouraged to raise a question of any sort in the video format and submit it through email, upload on the website (http://www.peopo.org/2008vote.php), or compress the file in CD and mail it to the organization in charge of this activity, which was *PeoPo*

Citizen Journalism Portal. The public was also encouraged to record the content using multimedia-enabled mobile phones and submit it though the mobile Web.)

Mobile phones and the other media can also act together to realign the technologies of control on individuals, meaning that people are offered the opportunity to observe those in authority, a concept often called "sousveillance" (Mann, Nolan, & Wellman, 2003). During the 2000 presidential election in Ghana, voters who were prevented from participating at the polls used mobile phones to make their experience publicly known by calling in to radio talk shows. These stations then broadcast the reports of voter intimidation, which in a way forced the police and the authorities concerned to embark on corrective actions to address these complaints. In the U.S., similar forms of sousveillance occurred when voters used camera-equipped mobile phones, along with blogs, to document evidence of voting irregularities during the 2006 congressional elections (Zuckerman, 2007).

The evidence sketched thus far helps bring to light a bigger picture under which mobile political campaigns are placed. Mobile phones, instead of being appropriated independently, need to be viewed as lodged within existing social, political and media contexts in contemporary political campaigns. As can be expected, the mobile will allow political campaigns to exploit its specific malleable characteristics, in conjunction with other ICTs.

Who Is in Control of Mobile Campaigns?

Online and mobile communication is endowed with the capacity to facilitate horizontal and decentralized networks that take autonomous forms (Castells, 2007). In the context of political campaigns, this advanced feature thus opens up the opportunity for people to seek and share information about candidates with ease. To some extent, these digitally traceable activities also help campaigns to record and possess personal information about the potential voters, which can be used to build an opt-in directory of supporters for mobilization purposes. The "double-edged" control issue is not unfamiliar in the discussion of mobiles communication (Geser, 2003). Yet, unlike the online environment where diverse forms of information flow are possible and traceable, mobile phones have been used mostly with the simple text-messaging format for political campaigns (and though often traceable, requires much more effort to effect). It therefore raises the question: Has the change of control in terms of information flow observed in the Internet arena (Howard, 2006) occurred in mobile-aided campaigns? What are the roles of candidates and citizens in these campaigns?

Campaigns' Reach and Coordination of Stakeholders. With the prevalence of digital media in political campaigns, citizenship is presumed to become "thinner" because people can become easily politically expressive without engaging themselves substantially in the process (Howard, 2006). For example, though it may not necessarily generate meaningful political discussion, a simple act of forwarding campaign messages through email or becoming friends with candidates on Facebook can be ways to participate in the campaign. To attract those emerging "thin" citizens' attention and retain their interest, campaigns regularly resort to various viral marketing strategies.

Note that the distinction has been made between *centralized viral political marketing*, which embodies a centralized mechanism through which parties or candidates consciously and professionally appropriate viral politics for their use, and *viral politics*, which underscores the horizontal communication among motivated citizens to voluntarily spread messages across the electronic networks (Dányi & Sükösd, 2003). Examples of the former include vote reminders through SMS, updates to supporters who choose to receive campaign news on their mobiles, or use of online media as an echo chamber to convey campaign messages to the mainstream media. The latter can be well explained by people's active efforts to reinvent and personalize the centralized produced campaign messages or mobilization through SMS networks. Weighing these two forms of viral political techniques, Dányi and Sükösd argue that the emerging tension manifests itself in the control crisis faced by campaign planners. That is, campaign planners are forced to enter an essentially uncontrollable space where citizens can personalize and mass circulate their own campaign-related messages just as easily and efficiently as the campaigns can distribute centrally produced messages.

In addition to different hierarchical approaches, one can see a progression in the scope of stakeholders campaigns can reach with the aid of mobile phones. In early adoptions of mobile phones in electoral campaigns, such as the 1997 U.K. elections, mobile phones were viewed by elite politicians as "parasitical" tools whose function was subsumed under the broader goal of more efficient use of the mass media or as a medium to maintain centralized control of the agenda among party managers who were geographically dispersed (Coleman, 1998, 1999). In Indonesia's 1999 general elections, mobile phones were used by the capital city politicians to organize political activities and used by leaders of parliamentary factions to recruit voting alliances on issues of interest (Hill, 2003).

But more recently, mobile technologies have been used by campaigns to reach and coordinate a wider scope of voters and supporters. In the 2001 U.K. elections, mobile communication was incorporated as part of the e-marketing efforts of the parties to achieve instant internal communication as well as external recruitment of voters and supporters (Coleman, 2001). In Kenya's 2002 elections, people in the remote areas were alerted to vote by the ruling KANU party operatives through SMS or mobile phone calls and were then transported by the provincial administration officers and civil servants to the polling stations (Otieno, 2005). In recent U.S. elections, the Republican Party's chief strategist, Karl Rove, coordinated the 2002 Republican Congressional victories in part through his BlackBerry-enabled networks (Carney & Dickinson, 2002). Democratic campaigners also used BlackBerries to coordinate get-out-the-vote efforts among volunteers for two House campaigns on Election Day (Krebs, 2002). On a peer-to-peer basis, Howard Dean's 2004 campaign used UPOC, a wireless community service, to communicate with and coordinate supporter groups (see http://www.upoc.com/group.jsp?group=dean-2004).

Citizens' Bottom-Up and Personalized Mobile Campaigns. The structural features (e.g., size limitations) are sometimes considered an obstacle for SMS to be used to effectively communicate with potential voters. Yet, paradoxically, these limitations can actually reinforce the personalized nature of campaign messages sent through mobile phones and may therefore increase the motivation of those who opt in to join the campaign activity (Hermanns, 2008). Specifically, jokes or poetic expressions tend to be more popular than hard facts or serious information and are thus more likely to be forwarded by mobile phone users (Dányi & Sükösd, 2003; Rafael, 2003). During Hungary's 2001 elections, it was not uncommon to see people create messages "worthy of forwarding, and passing on the ones made up by the parties—this represents a significant change compared to the unidirectional communication pattern of traditional mass media" (Sükösd & Dányi, 2003, p. 227). Likewise, during Kenya's 2002 elections, SMS messages usually travelled from the oppositional activists, who attempted to devise negative campaign messages about the ruling party, to opposition sympathizers, who then translated such message into their own vernaculars and passed it on to their social contacts (Otieno, 2005).

Much has been said about how political candidates and citizens utilize different methods to talk about and engage in campaigns. With the aid of mobile communication, candidates expand the area of interaction with their stakeholders while citizens capitalize on the opportunity to personalize the campaign messages as gatekeepers, consumers and producers (Hermanns, 2008). Interestingly, in the face of citizens' increasingly powerful private consumption of campaign messages, campaigns are struggling to stake out their controllable space by maintaining top-down communication patterns while also slowly increasing the level at which citizens, or at least supporters, can become involved.

How Does the Influence of Mobiles Campaigns Take Form?

The struggle to maintain control of the campaign speaks to a fundamental process of political communication: exerting influence on individuals in the form of attitude and behavior change. A critical aspect that merits attention is the role of informal social networks. The 2008 Obama campaign used the iPhone application "Call a Friend in a Swing State" to solicit supporters for local activities through their social networks. Unlike typical voter contact models, this network-based approach empowers users to call their personalized list of voters by sorting friends' phone numbers according to "key battleground states" in order to convey focused campaign messages (Melber, 2008). In the aforementioned model of decentralized viral politics, informal social networks are the route of transmission of political messages, through electronic networks as well as other informal methods of circulating political messages independently of the mainstream media (Dányi & Sükösd, 2003). In short, these practices go to the heart of the two-step flow of communication (Katz & Lazarsfeld, 1955), a notion that underscores the flow of influence passing from mass media to individuals through a social mediation process, typically initiated by opinion leaders of a social group.

The Role of Informal Networks. Building on the two-step flow model, informal networks enabled through mobile phones can be viewed as the mediator as well as the result of contemporary campaigns. In the 2002 Korean presidential election, mobile phones, as part of new network information technologies, were deemed critical to bringing together *ad-hoc* networks of younger voters to mobilize for the candidate they supported (Han, 2007). In Spain's 2004 general elections, SMS networks contributed to quick self-organized and unplanned demonstrations as overlapping social networks of contacts were tapped for mobilization in real time (Suarez, 2006). Otieno's (2005) study also showed that viral SMS campaigns dovetailed with the extended family-network structure in Kenya. In Kenya, rural-based family members are often included in the mobile phone subscription plans of their urban relatives. In the process of obtaining information about elections, the rural-based family members thus relied on their urban relatives to feed relatively accurate information through mobile phones. Other cases of campaigns reaching potential voters through a two-step flow of influence via their supporters on mobiles can be observed in recent elections in Thailand (Pongsawat, 2007) and the Philippines (Karen, Gimeno, & Tandoc, 2009).

There are claims that the social mediation process in political campaigns will be replaced by targeted marketing and narrowcasting because campaigns go to great lengths to use digital technologies to design and communicate well-tailored messages directly to individual voters (Bennett & Manheim, 2006). Nonetheless, more evidence suggests the opposite. In a sense, the role informal social networks play in contemporary campaigns resembles the traditional people-intensive form of earlier campaign communication, that is, direct interpersonal communication among candidates, volunteers and citizens at local levels (Norris, 2005). Through the use of the Internet and mobile phones, a person-focused pre-modern form of campaign seems likely to reappear (Sükösd & Dányi, 2003).

What Are the Outcomes of Mobile Campaigns?

Different forms of positive effects as a result of mobile phone use in political campaigns have been widely discussed, including mobilizing voter support, coordinating volunteers and facilitating transparent electoral procedures. Yet, inevitably, creative ways of harnessing mobile communication are followed by unintended consequences. In Kenya's 2007 elections, for example, hate speech circulated via SMS networks, raising tensions and ethnic hatred, which facilitated subsequent violence across the country (Abdi & Deane, 2008). In many cases, the unbridled practices of mobile communication essentially reflect the results of citizens being manipulated by political parties or those with vested interests (Osborn, 2008; Sükösd & Dányi, 2003).

From the perspective of the authorities and campaign planners, mobile communication allows for self-organized demonstrations and creative personalization of campaign messages, which may nevertheless pose threats to their control. This crisis of losing control often prompts government to initiate countermeasures or stockpile defensive reactions. For example, in Albania and Ethiopia, SMS networks were blocked by governments during or before elections (Zuckerman, 2007). Similar large-scale suspension of mobile networks by the government took place in Iran's 2009 election. There are also other recurring concerns expressed by governments concerning the trivializing effects of text messaging on the election process (Pratchett, 2002; Schwankert, 2008).

Conclusion

The questions sketched in this chapter outline the aspects critical to understanding how mobile communication has been appropriated in political campaigns, especially with variations under different social and political contexts. Certainly, there are other veins that demand further investigation, including how campaigns manage mobile communication across different stages of campaigning, and how the mobile loop (consisting of candidates and their supporters) co-evolves or interacts with the bigger decentralized network of a general public now equipped with online and mobile technologies. Though there are variations in different countries, people reached through mobiles are mostly those who choose to receive campaign messages and are somewhat favorable to the target candidates. The ways in which mobile communication, relative to the traditional and online media, can efficiently and effectively be used to mobilize supporters, attract potential voters or even formulate public opinion thus becomes a practical question to probe. At the societal level, a similarly pivotal question is how the mobile and media strategies transform themselves, or dissolve, after the campaigns end and the society moves into the next stage of public governance. Political campaigns are of social significance not merely because they stimulate the public or produce satisfying results. They also signal the inception of a feasible democratic process that can be put into practice based on the dynamics generated during the campaigns. As such, mobile campaigns will doubtless continue to play an ever-growing role in the way people participate in elections as well as seek to challenge or direct the governmental policy.

References

Abdi, J., & Deane, J. (2008, April). The Kenyan 2007 elections and their aftermath: The role of media and communication. BBC World Service Trust. Retrieved August, 31, 2008 http://downloads.bbc.co.uk/worldservice/trust/pdf/kenya_policy_briefing_08.pdf.

Adelman, J. (2004, July 12). U Say U want a revolution. Time Asia. Retrieved August 30, 2008, from http://www.time.com/time/asia/magazine/article/0,13673,501040712-660984,00.html.

Anstead, N., & Chadwick, A. (2009). Parties, election campaigning and the internet. In A. Chadwick & P. N. Howard (Eds.), Routledge handbook of Internet politics (pp. 56-71). New York: Routledge.

Bennett, W. L., & Manheim, J. B. (2006). The one-step flow of communication. The Annals of the American Academy of Political and Social Science, 608, 213-232.

Carney, J., & Dickinson, J. F. (2002, November 9). W. and the 'Boy Genius'. *Time*. Retrieved September 3, 2008, from http://www.time.com/time/nation/article/0,8599,388904,00.html.

Castells, M. (2007). Communication, power and counter-power in the network society. *International Journal of Communication, 1*, 238-266.

Coleman, S. (1998). The televised leaders' debate in Britain: From talking heads to headless chickens. *Parliamentary Affairs, 51*(2), 182-97.

Coleman, S. (1999). The new media and democratic politics. *New Media & Society, 1*(1), 66–78.

Coleman, S. (2001). Online campaigning. *Parliamentary Affairs, 54*, 679–88.

Dányi, E., & Sükösd, M. (2003). Who is in control? Viral politics and control crisis in mobile election campaign. In K. Nyíri (Ed.), *Mobile democracy: Essays on society, self and politics* (pp. 285–315). Vienna: Passagen Verlag.

Dale, A. L. & Strauss, A. B. (2007, April). *Text messaging as a youth mobilization tool: An experiment with a post-treatment survey*. Paper presented at the annual meeting of the Midwest Political Science Association, Chicago, USA. Retrieved August 30, 2008, from http://www.mindlessphilosopher.net/princeton/aapor_dale_strauss_2.1.07.pdf.

Geser, H. (2003). *Towards a sociological theory of the mobile phone*. Retrieved April 20, 2009, from http://socio.ch/mobile/t_geser1.htm.

Goldstein, J, (2007). The role of digital networked technologies in the Ukrainian Orange Revolution. *Berkman Center Research Publication No. 2007-14.*

Han, J. (2007). From indifference to making the difference: New networked information technologies (NNITs) and patterns of political participation among Korea's younger generations. *Journal of Information Technology & Politics, 4*(1), 57-76.

Hermanns, H. (2008). Mobile democracy: Mobile phones as democratic tools. *Politics, 28*(2), 74-82.

Hill, D. (2003, December). Communication for a new democracy: Indonesia's first online elections. *Pacific Review, 16*(4), 525-547.

Howard, P. N. (2006). *New media campaigns and the managed citizen*. New York: Cambridge University Press.

Karen, K., Gimeno, J. D. M., & Tandoc Jr., E. (2009). The Internet and mobile technologies in election campaigns: The GABRIELA Women's party during the 2007 Philippine Election. *Journal of Information Technology & Politics, 6*(3), 326-339.

Katz, E., & Lazarsfeld, P. F. (1955). *Personal influence: The part played by people in the flow of mass communications*. New York: Free Press.

Katz, James E., & Aakhus, M. (2002). Apparatgeist. In J. E. Katz & M. Aakhus (Eds.), *Perpetual contact: Mobile communication, private talk, public performance* (pp. 287-317). Cambridge, UK: Cambridge University Press.

Katz, James E. (2009). Media, la democrazia, e l'Amministrazione Obama: la speranza, senza cambiare? *Politica Communicazione, 10* (3), 421-31.

Kim, S. (2007, October 9). Sohn leads in mobile phone voting. *Korea Times*. Retrieved September 20, 2009, from http://www.koreatimes.co.kr/www/news/nation/2007/10/116_11575.html.

Krebs, B. (2002, October 11) Technology shapes get-out-the-vote efforts. *The Washington Post*. Retrieved November 30, 2005, from http://www.washingtonpost.com/ac2/wp-dyn?pagename=article&node=&contentId=A2467-2002Oct9&notFound=true.

Kreiss, D. (2009). Developing the "good citizen": Digital artifacts, peer networks, and formal organization during the 2003-2004 Howard Dean Campaign. *Journal of Information Technology & Politics, 6*(3), 281-297.

Lee, T. & Willnat, L. (2006) Media research and political communication in Singapore. *Working Paper No. 130, Asia Research Centre: Murdoch University, April.* Retrieved August 30, 2008, http://wwwarc.murdoch.edu.au/wp/wp130.pdf.

Mann, S., Nolan, J., & Wellman, B. (2003). Sousveillance: Inventing and using wearable computing devices for data collection in surveillance environments. Surveillance & Society 1(3), 331-355.

Melber, A. (2008, October 8). Obama's web-savvy voter plan. The Nation. Retrieved March 20, 2010, from http://www.thenation.com/doc/20081027/melber.

Mobiledia.com (2005, February). Text message reminder to vote in Thailand. Retrieved August 30, 2008, from http://www.mobiledia.com/news/25325.html.

Murphy, D. (2007, October 22). Once we were union, too. The Sydney Morning Herald. Retrieved August 30, 2008, from http://www.smh.com.au/news/federalelection-2007news/once-we-were-union-too/2007/10/21/1192940905333.html.

Norris, P. (2005). Political parties and democracy in theoretical and practical perspectives. Washington, DC: NDI. Retrieved August 30, 2008, from http://www.iknowpolitics.org/files/1950_polpart_norris.pdf.

Otieno, I. O. (2005, October). Mobile telephony and democratic struggles: A case of 2002 elections in Kenya. Paper presented at 'RE: activism: Re-drawing the Boundaries of Activism in New Media Environment. Centre for Media Research and Education, Department of Sociology and Communication, Budapest University of Technology and Economics, Budapest.

Osborn, M. (2008). Fuelling the Flames: Rumour and Politics in Kibera. Journal of Eastern African Studies, 2(2), 315-327.

Pongsawat, P. (2007). Middle-class ironic electoral cultural practices in Thailand. In B. H. Chua (Ed.), Elections as popular culture in Asia (pp. 94-138). New York and London: Routledge.

Pratchett, L. (2002). The implementation of electronic voting in the UK. Local Government Association. Retrieved September 1, 2008, from http://www.dca.gov.uk/elections/e-voting/pdf/e-voting.pdf.

Prete, M. I. (2007). M-politics: Credibility and effectiveness of mobile political communications. Journal of Targeting, Measurement and Analysis for Marketing, 16, 48-56.

Rafael, V. L. (2003). The cell phone and the crowd: Messianic politics in the contemporary Philippines. Public Culture, 15(3), 399-425.

Rajeev Gowda, M. V., Narayan, C. S., & Ollapally, J. (2006). Electoral reforms in India. In R. K. Mitro (Ed.), E-government: Macro-issues (pp. 24-44). New Delphi: GIFT Publishing. Retrieved August 30, 2008, from http://www.iceg.net/iceg2006gmi.pdf#page=31.

Schwankert, S. (2008, March 5). Chinese lawmaker condemns text message voting. ABC News.com. Retrieved August, 30, 2008, from http://abcnews.go.com/Technology/PCWorld/story?id=4393238.

Smith, A. (2009). The Internet's role in campaign 2008. Pew Internet and American Life Project. Retrieved March 20, 2010, from http://www.pewinternet.org/~/media/Files/Reports/2009/The_Internets_Role_in_Campaign_2008.pdf.

Steere, M. (2008, August 25). Cell phones promise fairer elections in Africa. CNN. Retrieved August 30, 2008, from http://www.cnn.com/2008/WORLD/europe/08/25/Cellphonedemocracy/.

Suarez, S. (2006). Mobile democracy: Text messages, voter turnout, and the 2004 Spanish elections. Representation, 42(2), 117-128.

Sükösd, M., & Dányi, E. (2003). M-politics in the making: SMS and e-mail in the 2002 Hungarian election campaign. In K. Nyíri (Ed.), Mobile communication: Essays on cognition and community (pp. 211-232). Vienna: Passagen Verlag.

Vargas, J. A. (2007, June 30). Text-friendly hopefuls vie for hearts and thumbs. Washington Post. Retrieved August 30, 2008, from http://www.washingtonpost.com/wp-dyn/content/article/2007/06/29/AR2007062902352.html.

Ward, S. J., & Gibson, R. K. (2003). On-line and on message? Candidate websites in the 2001 General Election. British Journal of Politics and International Relations, 5(2), 108-256.

Yeung, R. (2007). The emergence of new media in Hong Kong politics. Hong Kong Journal, 8. Retrieved August 30, 2008, from http://www.hkjournal.org/PDF/2007_winter/2.pdf.

Zuckerman, E. (2007). Mobile phones and social activism: Why cell phones may be the most important technical innovation of the decade. Techsoup. Retrieved August 30, 2008, from http://www.techsoup.org/learningcenter/hardware/page7216.cfm?cg=searchterms&sg=Zuckerman.

16

Disability, Mobiles, and Social Policy: New Modes of Communication and Governance

Gerard Goggin

People with disabilities have been avid users of mobile communications, using them in their everyday lives—in prosaic as well as novel ways. Where possible, people with disabilities took up relatively bulky first-generation cell phones, then second-generation and third-generation devices after that, welcoming them for the communication possibilities they offered over and above the limited of fixed-line telephone. William Jolley, for instance, notes that:

> People who are blind have enjoyed the flexibility that results from mobile commu-
> nications. Using their mobile phones they can find each other more easily in public
> places, and they have the added security of being able to make a phone call if they
> are lost or feeling endangered (Jolley, 2003, p. 27).

For deaf people, even first-generation mobile phones were important:

> Analogue-based mobile phones emerged ... in the 1980s, and while designed
> primarily for voice communication, some models did work with portable TTYs
> [teletypewriters—text communication devices used by Deaf people] ...This further
> improved the ability of Deaf people to make remote contact with people while on
> the move (Harper, 2003, p. 155).

> Then text messaging capabilities of second-generation cell phones proven an even
> greater boon (Power & Power, 2004), with third-generation promising, if not entirely,
> delivering the much desired video technology that would allow communication in
> sign language (Harper, 2003).

Sometime in 2010, the number of mobile phone subscribers world-wide topped the five billion mark. While there are no accurate figures on the number of people with disabilities using mobiles, it is helpful to consider general statistics on disability. Estimates vary, but people with disabilities are believed to number up to twenty percent of the popula-tion in many countries (especially developed countries such as US,

Canada, and Australia). For various reasons developing countries report lower prevalence of disability, and so "a worldwide estimate of about a 10-12 percent rate of disability seems reasonable" (World Bank, 2009; see also Mont, 2007). There is an "intricate" link between disability and poverty, with people with disabilities being "among the poorest of the poor" (Mont & Loeb, 2008). Bound up with the threshold issue of classifying and counting disability is its very definition. Approached in quite different ways across countries and cultures, nonetheless it is fair to say that disability has been conceived on what has been termed the "biomedical" model, or via key discourses such as that of charity (Fulcher, 1989; Goggin & Newell, 2005). This matters because the alternative approaches emphasize the social, political, and cultural de-termination of disability (Campbell, 2009; Shakespeare, 2006; Siebers, 2008). For instance, the "social model," originated in Britain, holds that disability is the social production of impairment (Oliver, 1996). That is, people may be born with, or acquire, impairments. However, it is society and its barriers to participation that very often disables them, by discrimination, exclusion, and poor design. I mention these important, ongoing debates about disability because they go directly to the heart of we see disability in society—and, consequentially, how we construe required social policy when it comes to mobile communication.

There is a range of pressing social policy issues that flow from dis-ability, however regarded. The most general statement of this in recent times may be found in the United Nations Convention on the Rights of Persons with Disabilities (CRPD), which came into force, along with its optional protocol, on May 3, 2008. As the UN explains:

> The Convention marks a "paradigm shift" in attitudes and approaches to persons with disabilities. It takes to a new height the movement from viewing persons with disabili-ties as "objects" of charity, medical treatment and social protection towards viewing persons with disabilities as "subjects" with rights, who are capable of claiming those rights and making decisions for their lives based on their free and informed consent as well as being active members of society. The Convention is intended as a human rights instrument with an explicit, social development dimension (UN, 2008).

The Convention is undergoing the long process of ratification, and countries giving effect in their national laws and practices to its provi-sions. What is notable, however, is its many provisions that concern mobiles and associated technologies, especially the Internet. This con-vention is perhaps the first international treaty that recognizes the critical place that the mobile phone—and Internet, communications and media technologies—play in social life and social policy.

Many people have been placed in a double bind when it comes to mobiles. As we have seen, as the technology diffused through societies, people with disabilities, like many others, wished to avail themselves of mobiles. However they were restricted from doing so, by inaccessible devices and poorly designed interfaces, features, and applications (Bryen et al., 2005; Goggin & Newell, 2003, & 2006; Roe, 2001, & 2007; Shipley & Gill, 2000). The most striking recent example, in a long line of inaccessible mobile devices, is the glorious iPhone. Hailed as the "Jesus" phone, acclaimed for its cool design, as well as opening up the locked-down mobile phone platform, and so ushering in a new era of developer and user innovation, it took Apple at least two years to take accessibility for users with disabilities seriously. For many blind users and those with mobility impairments, it was not until the introduction of its VoiceOver interface in June 2009, that at least some of the functionality taken as standard in mobile phones and computers (via screen readers, for instance) was restored (Goggin, 2010; Lance, 2008; Mosen, 2009).

Despite this frustrating, and ongoing, struggle with accessibility, people with disabilities have keenly consumed mobiles. In the process they have developed instinctive and innovative cultures of use, and devised creative ways to domesticate the technology. Two examples will suffice here. Firstly, that of blind users, who use screen reader software to navigate and customize mobile phones (Goggin & Newell, 2006), and are now investigating new possibilities of mobile media, not least GPS and locative media. Second, the case of deaf users, who, among various applications have created new possibilities for text messaging (Bakken, 2005; Harper, 2002; Pilling & Barrett, 2007; Power & Power, 2004; Power, Power, & Rehling, 2007; Valentine & Skelton, 2008 & 2009) and also pioneered sign-language video mobile telephony (Cavender et al., 2008; Hilzensauer, 2006). Though further research is certainly needed, there are clear signs that mobile communication plays a vitally important role in the lives of people with disabilities—in ways which are still to be generally appreciated.

These introductory remarks—distinctive forms of mobile communication inaugurated by people with disabilities, and social policy as it relates to disability—are important to set the scene for what follows: a critical discussion of the mobile communication-social policy nexus as it relates to people with disabilities. To undertake this, firstly, the chapter discusses key issues in the development of social programs for people with disabilities that are delivered via mobiles. Secondly, the chapter discusses case studies where mobile technology has been used in campaigning and advocacy by people with disabilities to influence social policy. Thirdly, I look at new mobile communication technologies, such as Twitter, and the emerging kinds of social policy formations around disability these are supporting.

My argument is that new modes of communication and governance are emerging in social policy with respect to disability. People with disability, their allies, and others are beginning to create and secure new social policy concepts and programs, in which mobile communication technologies are playing an important part. Further, new modes of governance are now possible, that draw upon the affordances and cultures of mobile and online technologies, as well as the new understanding of people and their lives that goes along with a human rights approach to disability.

Social Programs, Disability, and Mobiles

Since the spread of mobiles in the mid-to-late 1990s, the technology has featured in a range of social programs. These vary widely from the provision of essential information and emergency services, through health and wellbeing, and employment, to initiatives promoting self-esteem and social inclusion. People with disabilities have often been the target of these programs. While it is not possible to fully discuss these here, some remarks are in orders. Often these social programs devised for delivery with mobile communication are articulated with the goal of ameliorating, or overcoming, the perceived challenges faced by people with disabilities. There is a consistent view that mobiles are important for the welfare, well-being, and inclusion of people with disabilities in society (Keating, 2007). The project of providing people with the support for greater independence or autonomy in their lives is a key goal for which mobile phones are typically marshaled. We find statements, for instance, along the lines that "user groups with mobility restrictions is one of the human groups for which the mobile phone can be beneficial" (Abascal & Civit, 2000; cf. Imrie, 2000). Also commonly encountered is the idea of providing portable, personal support, in situ, for people facing particular challenges, through their impairments or disabling situations. One example of this is a project that used the mobile Internet to deliver information in real-time to individuals with autistic spectrum disorders and social phobia, while they were in social situations they found difficult or challenging (Bishop, 2003). With the advent of smartphones and mobile media, there are even richer and more ingenious possibilities for social policy programs that address key aspects of people's lives and their welfare. To mention just a few: the possibilities of independent navigation and way-finding for blind people or people with cognitive impairments, using localization and positioning technologies with mobiles (Liu et al.,

2007); mobiles as memory aids (Caprani et al., 2006; Stapleton et al., 2007); the sensing and sensory capability of mobiles (Ipiña et al., 2005); or mobile learning for people with disabilities (Revelle et al., 2007).

As this brief survey indicates, there is a welter of programs that deploy mobiles to address aspects of social policy for people with disabilities. One difficulty this poses for analysis of mobile communication, however, is that the social policy objectives in much of this are not especially clear. This is as much to do with a lack of clarity regarding the nature and characteristics of disability, as it is to do with mobile communication. Let me frame this in another way: many social programs for people with disability are on the cusp of different ideas of what disability subtends, or touches upon: health; welfare; impairment; aging; dependence versus independence (rather than, perhaps, interdependence); social isolation; language and literacy. Perhaps this is not such a problem, given disability is a large and complex topic, requiring a multitude of coordinated policy objectives and programs. However, I wonder if there are unhelpful assumptions about disability being reproduced, rather than challenged. The importance of contesting and indeed transforming such disabling assumptions, lies in the potential that mobiles could then offer to open up new directions for social policy and disability.

Here a comparison might be helpful. There is a burgeoning area of innovative programs that utilize mobiles, especially for community, social, and economic development—notably articulated towards the booming markets in mobiles in the developing world (Donner, 2007; Goggin & Clark, 2009; Hughes & Lonie, 2007). Such directions in mobiles and social policy have been promoted by international agencies, such as the UN and World Bank, in conjunction with major mobile companies such as Vodafone (for instance Kinkade & Verclas, 2008). There are good reasons to critically evaluate as well as engage with these directions in mobiles, however one detects a quite different flavor to these discourses that in those concerning mobiles and disability. There is an emphasis on mobiles and social entrepreneurship, for instance—as in the celebrated case of the Bangladeshi Grameen Bank (Moni & Uddin, 2005; Sullivan, 2007)—that recognizes and seeks to capitalize on the economic and social development of the individuals and communities framed as subjects of the policy that does not often figure when it comes to disability. My hunch is that this is because the new approaches take disability as generative—rather than a deficit that requires special, post-hoc, treatment—are not widely understood or credited. To seek such empowering, affirmative models of disability and mobiles for social policy, I want to turn now to explore disability intervention into social policy using mobiles.

Disability Advocacy and Campaigning

Because of their specific uses and affordances for people with disabilities, mobiles have taken on a significant role in advocacy and campaigning associated with social policy. There is, however, still little explicit, detailed research regarding this. The area of social policy in relation to technology is still an overlooked one—often regarded as a specialist or technical area. With the rise of digital technologies, and the enshrining of their importance in all facets of business, education, politics, cultural and social life, social policy matters have gained salience and significance. While there have many longstanding social policy issues associated with telecommunications and computing—affordability, access, universal service, consumer protection and information, privacy, to mention a few—it was not until the liberalization of markets, and introduction of new technologies, through the 1980s and 1990s that these were given sustained, comprehensive attention. From the late 1980s onwards, mobile communication began to emerge as an important topic of social policy, and one that raised new issues for policymakers. Alongside mobiles, and increasingly entwined with them, was the fast rising Internet. Where the issues of digital technologies and society irrupted onto the world stage, for the consideration of wider audiences and publics was the World Summit of the Information Society (WSIS).

WSIS comprised a preparatory process and two phases, held in Geneva, in December 2003, and in Tunis, in November 2005. What was notable about WSIS for those interested in communications and policy was the unprecedented extent to which civil society, especially through non-government organizations (NGOs), maneuvered their way to centre-stage of international, state-based discussions on the future of digital technologies. The other thing worth remarking about WSIS is the extent to which it marked a new era in the use of digital technologies themselves to network among the NGOs particularly, dispersed as they were across the world, and lacking resources for extensive face-to-face meetings and other resource-intensive forms of collaboration. The Internet especially became a key to this new form of campaigning. Further, the Internet itself generated new, decentralized and dispersed models for conceiving policy processes and effecting policy goals. All this is the subject of a considerable body of research (Raboy & Landry, 2005; Servaes & Carpentier, 2006). What does not make much of an appearance in this literature are two things: the emergence of disability as a shared, supra-national concern and constituency in social policy conceived via digital technology; and the use of mobile communications in the organizing among activists and citizens that resulted in the emergence of specific disability claims regarding the character and future of the world information society.

With the UN Convention on the Rights of Persons with Disabilities, a new era in social policy dawned. At roughly the same time, the use of mobiles in WSIS, becomes prominent in disability activism (on changing notions of disability activism and global policy, see Kitchin & Wilton, 2003; and Priestly, 2007). A feted instance of mobiles used for disability activism occurred in the revocation of the appointment of Jane K. Fernandes, the incoming president of the famous Gaulladet University—the premier university for the deaf in the world.

Gaulladet students are renowned for their political commitment and acumen, shown in the long-running protest that led to the appointment of the first deaf president of the university, I. King Jordan. As the replacement for Jordan when he retired, the board of trustees chose Fernandes, the former provost. There was widespread criticism, however, and after six months of protests, the board gave in and revoked Fernandes' appointment (Goldenberg, 2006; De Vise, 2009). Key to the students' success in mobilizing protest, if press reports are to be credited, was their use of mobile communication:

> The successful mobilization can be partly credited to a technology that did not exist a decade ago: the wireless handheld computer. For a month, protesters used the mobile devices to wage a wireless war via messages to each other, the media and the international deaf community. Almost every Gallaudet student has a BlackBerry, Sidekick or other handheld. The students say the technology has brought them more equality and has opened up the world. From their mobile devices, protesters say they sent batch e-mails with information and updates to large lists of e-mail addresses. They also instant-messaged fellow protesters to tell them where to go for impromptu rallies … (Takruri, 2006).

Figure 16.1 "Al Jiminez, a senior at Gallaudet University, uses a wireless handheld computer to communicate with other student activists" (Takruri, 2006)

In covering the Gaulladet protests, the reporter evoked two well-known exemplars of mobile activism (Takruri, 2006): the downfall of Philippines President Joseph Estrada in the famous "coup de text" (Goggin, 2006; Pertierra et al., 2002); and Howard Rheingold's "smart-mobs" (Rheingold, 2002). This genealogy can be seen in more recent causes célèbres of mobile activism, such as the protests in Burma or the "green revolution" in Iran in 2009. The difficulty, which at least the journalist is aware of, is specifying the role that technology actually did play in the processes of organizing, politics and policy (Castells, 2009; Goggin, 2006).

In the case of the 2006 Gaulladet protests, there is, of course, an obvious historical comparison to assist us: the 1988 "Deaf President Now" protests that led to I. King Jordan's appointment. In discussing why the 1988 protest was so successful, the Gaulladet website suggests, among other factors, that:

> The protesters used the same methods—boycotting of classes, marches, and letter writing campaigns that had historically been successful strategies of protests in the past ... There was phenomenal media attention and coverage during the entire week ... It was one of the first times for the reporters and the viewers alike to see for themselves that deaf students and deaf people really could do anything, except hear (DPN, 1998).

The role that a range of media played was pivotal also in the 2006 dissent event. This did include television and newspapers, but also the interaction of these with the new media, especially developed by cultures of deaf use (Hogg et al., 2008; Read, 2006). For instance, the new interim president of Gaulladet, Robert Davila, employed video blogs to communicate with the students and university community:

> Davila started regular video blogs, or vlogs, in sign language to update the campus on matters such as accreditation ... Almost no other university leader is being watched so closely and judged so much—by federal agencies, accreditors and by the deeply engaged and ever-blogging deaf community (Kinzie, 2007).

Gaudallet plays an important role in public discussions of deaf cultures, their changing identities, and relations with non-deaf, hearing cultures (Leigh, 2009). A profile of Gaudallet's hearing football coach, Ed Hottle, emphasized how his successful rapprochement with his initially wary deaf players lay in his willingness to learn American sign language—but crucially also their mobile technology of choice:

> Hottle began using a hand-held BlackBerry-like device that no Gallaudet student is without. It's the deaf person's cellphone, allowing instant communication through text messaging. Before long, Hottle's wife had to beg him to put it away. At all hours of the night, his players would text him notes or tidbits about their lives—one explained

why he was sluggish in practice; another shared good news about his grades—and Hottle always responded. He knew it was an important way to get to know the players and gain their trust (Foster, 2007).

Others, such as Gaulladet professor Kathleen M. Wood, also saw the mobile as an iconic representation of the diversity yet particularities of deaf culture:

> The protesters know that removing Ms. Fernandes is only a first step toward creating a Gallaudet for the deaf students already here and in our future—students of diverse backgrounds with clear speech, eloquent ASL, cochlear implants and Sidekick pagers. The new Gallaudet will not be for out-of-step leaders or people who think that monolingualism and membership in a single culture are badges of honor. The new Gallaudet will not pretend to be for everyone (Wood, 2006).

The 2006 Gallaudet protests are an excellent example of the new modes of mobile communication being devised by people with disabilities—in this case, the deaf community—to intervene in policy processes. This case brings together in a highly visible case (if one still not sufficiently analyzed) an example that illustrates how people with disabilities may reshape the contours of policy arena. This arena may not be what is traditionally regarded as social policy, although of course, education is a key to life chances and outcomes. However, what it reveals is that we reframe disability as we come to recognize and respond to the new affordances provided by mobile communication. This in turn this can lead us to rethink social policy.

If it is true that mobile communication are important in disability and social policy programs, and also playing a nascent role in the organizing around policy, then it is also the case that such trends are deepening in the new phase of mobile media technologies. With mobile media, there is a deepening of the cross-overs between mobiles and the Internet, and also among different forms of audio, video, visual, and textual media—all of which brings mobiles into play as a potential open and configurable computing platform.

One example of an emerging formation at the interface of mobile communication and social policy may be found in Twitter, the short messaging application that already in its short career has proven a flexible and highly popular medium. Twitter brings together (in a new form) attributes of text messaging, instant messaging, email, micro and mobile-blogging, really simple syndication (RSS) feeds, just to mention a few prior technologies. Twitter has already attracted many users with disability, and is offering new ways to construct disability communities, as well as relations with non-disabled users. Twitter builds upon the already well-established cultures of text, audio, and video blogging among disability communities (Keating & Mirus, 2004).

A number of disability advocacy and information organizations now include Twitter among their embrace of new technology options. Publications now routinely use Twitter as part of their communication with readers. Examples include: DAA News, from the International Disability and Human Rights Network; the BBC's *Ouch!* (a longstanding innovator in online technologies); *Disabled World*; the UK *Disability Now*. Numerous disability activists, leaders, and simply people wishing to include disability as part of their diverse identity, are tweeting: Susan Fitzmaurice, S.E. Michigan ("Diehard disability activist"); Blind audio and Internet radio broadcaster and blogger, Jonathon Mosen, from Wellington, New Zealand. The Queensland Youth Affairs Network is not alone in marshalling support for a project, in this case, a Youth Disability Advocacy Service, via Twitter (Youth Affairs Network, 2009).

Crucially also, because of its role in broadcasting news, information, and links, Twitter is poised to play a unique role in social policy processes. Of course, Twitter is probably still predominantly a computer-based technology, but with smartphones, mobile email, and especially iPhones, Twitter bridges a number of technologies. As well as Twitter being a way for people with disabilities to gain new voice in social policy, it might also contribute to the displacement of the politics of voice and representation to that of listening (Crawford, 2009). That is, Twitter offers a way for people with disabilities to practice new techniques of listening that have potentially important implications for how policy-making takes place.

What is likely to emerge, though at the time of writing it does not seem to have been widely deployed, is the use of Twitter in what has been termed electronic consultation and government—perhaps blurring the boundaries between this and what is regarded as mobile ("m-") government. While there has been much work on disability and consultation, there has been little systematic discussion of disability and online modes of consultation, let alone mobiles and consultation—or the role of mobiles in facilitating citizen participation in policy and government. Discussing the Canadian experience, Stienstra & Troschuk note that "eConsultations are increasingly used by governments to engage citizens in the policy processes," and suggest that:

> eConsultations need to be inclusive; that is, work proactively to include often marginalized populations ... eConsultations by themselves will not reach all people, but together with more traditional forms of consultation, they may be able to increase

the possibilities for participations of all Canadians, including those with disabilities (Stienstra & Troschuk, 2005).

There has been some acknowledgment of disabled citizens, and accessibility issues, in the fledgling mobile government literature. However, such discussions are yet to incorporate the lessons from e-government and disability, let alone reflect upon the possibilities of mobile media and disability that Twitter and other mobile communications represent.

Conclusion

In this chapter, I have sought to open up the topic of mobile communication and disability and the dimension of social policy inherent in this. To do so, I started with a discussion mobile communication and disability, on the one hand, and disability and social policy, on the one other. It is by returning to first principles in both these areas, then tracing and redrawing new connections, that new, dynamic, and productive social policy approaches can be forged.

While there are many worthwhile programs in existence, underway, or projected that draw upon the capabilities, features, computing power, extensibility, or intimate nature of mobile devices to address social policy concerns, my concern is that many of these endeavors—perhaps the majority—fail to go to the heart of contemporary disability, and the discrimination, exclusion, and power relations that constitute it. This is where activism by people with disabilities, using mobiles, has much to teach us. It offers different images of social policy and mobiles. It also conjures up visions of how such new forms of communication might help us to create new modes of governance. Such political departures have been prefigured, and indeed materialized in different settings, from the decentralized, international arrangements of WSIS that are now working their way out in the contentious Internet Governance Forum, and also through the enormous deployment of online and mobile activism represented by the drafting and passing into international law of UN Disability Convention. New possibilities are unfolding in the new mobile-Internet hybrid technologies of Twitter, iPhone and smartphone apps, mobile Internet, mapping, location, and sensing technologies. In all of this, the possibilities are far too rich and exciting for the field of mobile communication and social policy, to allow narrow notions of disability to constrain new visions.

References

Abascal, J., & Civit, A. (2000). Mobile communication for people with disabilities and older people: New opportunities for autonomous life. In P. L. Emiliany & C. Stephanidis (eds.), *Information Society for All* (255-268). 6th ERCIM Workshop on User Interfaces for All. Firenze: Consiglio Nazionale delle Ricerche.

Bakken, F. (2005). SMS use among Deaf teens and young adults in Norway. In R. Harper, L. Palen & A. Taylor (eds.), *Inside text: Social, cultural and design perspectives on SMS* (pp. 161-174). London: Springer.

Bishop, J. (2003). The Internet for educating individuals with social impairments. *Journal of Computer Assisted Learning, 19*, 546-556.

Bryen, D.N., Carey, A., & Friedman, M. (2005). Cell phone use by adults with intellectual disabilities. *Intellectual and Developmental Disabilities, 45*(1), 1-2.

Campbell, F. (2009). *Contours of ableism: Territories, objects, disability and desire.* London: Palgrave Macmillan.

Caprani, N., Greaney, J., & Porter, N. (2006). A review of memory aid devices for an ageing population. *PsychNology Journal, 4*(3), 205-244.

Castells, M. (2009). *Communication power.* New York: Oxford University Press.

Cavender A., Vanam R., Barney D.K., Ladner R.E., & Riskin E.A. (2008). MobileASL: Intelligibility of sign language video over mobile phones. *Disability and Rehabilitation. Assistive Technology, 3*(1), 93-105.

Crawford, K. (2009). Following you: Disciplines of listening in social media. *Continuum, 23*(4), 523-535.

Donner, J. (2007). The use of mobile phones by microentrepreneurs in Kigali, Rwanda: Changes to social and business networks. *Information Technologies and International Development, 3*(2): 3-19.

Foster, B. (2007, August 19). Sound and the fury. *Washington Post,* W26.

Fulcher, G. (1989). *Disabling policies?: A comparative approach to education, policy, and disability.* London and New York: Falmer Press.

Goggin, G. (2010). *Global mobile media.* London and New York: Routledge.

Goggin, G., & Clark, J. (2009). Mobile phones and community development: A contact zone between media and citizenship. *Development in Practice, 19*, (4-5), 585-597.

Goggin, G., & Newell, C. (2003). *Digital disability: The social construction of disability in new media.* Lanham, MD: Rowman & Littlefield.

Goggin, G., & Newell, C. (2005). *Disability in Australia: Exposing a social apartheid.* Sydney: University of New South Wales Press.

Goggin, G., & Newell, C. (2006). Disabling cell phones: Mobilizing and regulating the body. In P.K. Anandam & N. Arceneaux (eds.), *The cell phone reader: Essays in social transformation* (pp. 155-72). New York: Peter Lang.

Goldenberg, S. (2006, October 31). "Not deaf enough" university head is forced out. *Guardian,* 22.

Harper, P. (2002). Networking the Deaf nation. *Australian Journal of Communication, 30*(3), 153-66.

Hilzensauer, M. (2006). A way of integrating Deaf, hearing- and speech-impaired people into modern communication society. *Journal of Telecommunications and Information Technology,* 2, 32-40.

Hogg, N. M., Lomicky, C. S., & Weiner, S. F. (2008). Computer-mediated communication and the Gallaudet university community: A preliminary report. *American Annals of the Deaf, 153*, 89-96.

Hughes, N. and Lonie, S. (2007) M-PESA: mobile money for the "unbanked" turning cellphones into 24-hour tellers in Kenya. *Innovations, 2*(1-2), 63-81.

Imrie, R. (2000). Disability and discourses of mobility and movement. *Environment and Planning A, 32*(9), 1641-56.

de Ipiña, D. L., Vázquez, I., & Sainz, D. (2005). Interacting with our environment through sentient mobile phones. Proceedings of Second International Workshop on Ubiquitous Computing (IWUC 2005), Miami, May 24-25 (pp. 19-27).

Jolley, W. (2003). *When the tide comes in: Towards accessible telecommunications for people with disabilities in Australia*, Sydney: Human Rights and Equal Opportunity Commission. Retrieved from http://www.hreoc.gov.au/disability_rights/communications/tide.htm.

Keating, E. (2007). *The role of the mobile phone in the welfare of aged and disabled people*. Austin: University of Texas; Tokyo: DoCoMo. Retrieved from www.moba-ken.jp/wp-content/pdf/finalreport_06-05.pdf.

Keating, E., & Mirus, G. (2004). American Sign Language in virtual space: Interactions between deaf users of computer-mediated video communication and the impact of technology on language practices. *Language in Society, 32*(5), 693-714.

Kinkade, S., & Verclas, K. (2008). *Wireless technology for social change: Trends in mobile use by NGOs*. Washington, DC, and Berkshire, UK: UN Foundation-Vodafone Group Foundation Partnership.

Kinzie, S. (2007). A Sense of urgency at the top. *Washington Post*, B03.

Kitchin, R., & Wilton, R. (2003). Disability activism and the politics of scale. *The Canadian Geographer, 47*(2), 97-115.

Lance, G.D. (2008). Forget the iPhone: Accessibility trumps trendiness. Retrieved from http://www.computingunplugged.com/issues/issue200803/.

Leigh, I. (2009). *A lens on Deaf identities*. New York: Oxford.

Liu, A.L., Hile, H., Kautz, H., Borriello, G., Brown, P.A., Harniss, M., & Johnson, K. (2007). Indoor wayfinding: Developing a functional interface for individuals with cognitive impairments. *Disability and Rehabilitation: Assistive Technology, 3*(1-2), 69-81.

Moni, M. H. & Uddin, M.A. (2005). Cellular phones for women's empowerment in rural Bangladesh. *Asian Journal of Women's Studies, 10*(1), 70-89.

Mont, D. (2007). *Measuring disability prevalence*. Social Protection discussion paper No. 0706. Washington, DC: World Bank. Retrieved from http://www.worldbank.org/disability.

Mont, D., & Loeb, M. (2008). *Beyond DALYs: Developing indicators to assess the impact of public health interventions on the lives of people with disabilities*. Social Protection discussion paper No. 0815. Washington, DC: World Bank. Retrieved from http://www.worldbank.org/disability.

Oliver, M. (1996). *Understanding disability: From theory to practice*. Houndsmill, Basingstoke: Macmillan.

Pertierra, R., Ugarte, E.F., Pingol, A., Hernandez, J. and Dacanay, N.L. (2002). *Txt-ing selves: Cellphones and Philippine modernity*. Manila: De La Salle University Press. Retrieved from http://www.finlandembassy.ph/texting1.htm.

Pilling, D., & Barrett, P. (2008). Text communication preferences of Deaf people in the United Kingdom. *Journal of Deaf Studies and Deaf Education, 13*(1), 92-102.

Power, M., & Power, D. (2004). Everyone here speaks TXT: Deaf people using SMS in Australia and the rest of the world. *Journal of Deaf Studies and Deaf Education, 9*(3), 333-343.

Power, D., Power, M., & Rehling, B. (2007). German Deaf people using text communication: Short Message Service, TTY, relay services, fax, and e-mail. *American Annals of the Deaf, 152*(3), 291-301.

Priestly, M. (2007). In search of European disability policy: Between national and global. *Alter: European Journal of Disability Research, 1*(1), 61-74.

Raboy, M., & Landry, N. (2005). *Civil society, communication, and global governance: Issues from the World Summit on the Information Society.* New York: Peter Lang.

Read, B. (2006, November 10). Technology and influential blogs helped galvanize protests at Gallaudet. *Chronicle of Higher Education, 53*(12). Retrieved from http://chronicle.com/article/TechnologyInfluential/16859.

Revelle, P., Reardon, E., Green, M.M., Betancourt, J., & Kotler, J. (2007). The use of mobile phones to support children's literacy learning. In Y. de Kort et al. (eds.), *Persuasive Technology* (pp. 253-258). Berlin: Springer-Verlag.

Rheingold, H. (2002). *Smart mobs: The next social revolution.* Cambridge, MA: Perseus.

Roe, P. (Ed.). (2001). *Bridging the gap? Access to telecommunications for all people.* COST291bis. Brussels: Commission of the European Communities. Retrieved from http://www.tiresias.org/phoneability/bridging_the_gap/index.htm.

Roe, P. (Ed.). (2007). *Towards an inclusive future: Impact and wider potential of information and communication technologies.* Brussels: COST219ter. Retrieved from http://www.tiresias.org/cost219ter/inclusive_future/index.htm.

Servaes, J., & Carpentier, N. (eds.). (2006). *Towards a sustainable information society: Deconstructing WSIS.* Bristol: Intellect.

Shakespeare, T. (2006). *Disability rights and wrongs.* London and New York: Routledge.

Shipley, T., & Gill, J. (2000). *Call barred? Inclusive design of wireless systems.* London: PhoneAbility. Retrieved from http://www.tiresias.org/phoneability/wireless.htm.

Siebers, T. (2008). *Disability theory.* Ann Arbor: University of Michigan Press.

Stapleton, S., Adams, M., & Atterton, L. (2007). A mobile phone as a memory aid for individuals with traumatic brain injury: A preliminary investigation. *Brain Injury, 21*(4), 401-411.

Stienstra, D., & Troschuk, L. (2005). Engaging citizens with disabilities in eDemocracy, *Disability Studies Quarterly, 25*(2). Retrieved from http://www.dsq-sds.org/article/view/550/727.

Sullivan, N. P. (2007) *Can you hear me now? How microloans and cell phones are connecting the world's poor to the global economy,* San Francisco: Jossey-Bass.

Takruri, L. (2006, Nov. 8). Handhelds aided hearing-impaired protestors. *MSNBC.* Retrieved from http://www.msnbc.msn.com/.

United Nations (UN). (2008). Convention on the rights of persons with disabilities. Retrieved from http://www.un.org/disabilities/.

De Vise, D. (2009, September 5). Gallaudet University announces four finalists for President. *Washington Post,* B08.

Valentine, G., & Skelton, T. (2009) An umbilical cord to the world: The role of the internet in D/deaf people's information and communication practices. *Information, Communication and Society, 12,* 44-65.

Valentine, G., & Skelton, T. (2008). Changing spaces: The role of the internet in shaping deaf geographies. *Social & Cultural Geography, 9,* 469-485.

Wood, K. (2006, November 2). The right future for Gallaudet. *Washington Post,* A16.

World Bank. (2009, August 20). How many disabled people are there world wide? Retrieved from http://www.worldbank.org/disability.

Youth Affairs Network. (2009, February 3). Use your Twitter account to promote youth disability advocacy. Retrieved from http://www.yanq.org.au/content/view/1673/95/.

17

Social Participation and Mobile Communication

Leopoldina Fortunati and Anna Maria Manganelli

To date there have been several studies analyzing the political impli-cations of mobile phone use. These include the work of Katz & Aakhus (2002) on perpetual contact, Fortunati's (2002a) work on the mobile phone and its social role to enhance democracy in the communication environment, the seminal work of Rheingold (2002) on smart phones, an important book edited by Nyíri (2003) with a large section on mobile communication and democracy, the more recent efforts made by Castells and his colleagues on the relationship between mobile communication and globalization (2006) and by Wellman and Hogan (2006) on con-nected lives. Additionally, Fortunati's works on the use of Information and Communication Technologies (ICTs) as immaterial labor (2006, 2007) as well as the works of Ling (2008) on social cohesion and Ling and Donner (2009) on power and mobile communication help to shed light on the political meaning of mobile communication. There has also been a growing interest within the scientific community towards the use of mobile technology by public administration (Vincent, Harris, 2008) and other studies have focused on mobile communication and disability (See Goggin's chapter in this volume).

What remains to be explored is if and how mobile technology influ-ences the social policy environment. The research we present here does not directly analyze the relationship between social policy and mobile communication, but rather the relation between social participation and mobile communication; it is assumed that social participation is an influencing factor of social policy.

Elsewhere Fortunati analyzed the notion of social participation (in press), coming to the conclusion that participation is doubly ambivalent. It is ambivalent because it has a strong and a weak meaning (Gallino, 1993), but also because it presents positive as well as negative features.

In the strong sense, participation means the intervention of an individual or group on the governance of a community. In particular, it is a notion that refers to the concept of "participatory, direct democracy," in which all citizens should be actively involved for the important decisions. One of its practical limits is that participation slows down and complicates the decision-making process and can be really effective only in groups with 500 active members at most (Scott and Marshall, 2005).

In the weak sense of the term, participation means taking part to a greater or lesser degree in the activities of voluntary associations, groups of interest or religious groups. All these forms of participation are generally performed regardless of the possibility for the subject to influence the decisions that are made inside the governance of the community. The purpose of participation in these cases is to do something together for the community and/or for oneself.

In addition to its strong and weak meanings, the notion of participation is also ambivalent since, as is emerging from a number of research projects (Le Bouedec, 1984; Sarrica et al., 2009), participation is characterized both by the capacity to include in the community and to exclude from social life. In particular, the research carried out by Sarrica et al. (2009) on social representations of participation shows that for the group of young people investigated participation contains the germ of social exclusion and discrimination. Respondents denounce the sense of failure they feel when the contribution of each member is not appreciated in the same way or when the competition intervenes and there is a winner and all the others are excluded. Another research carried out by Farinosi (2010) on privacy, control and UGC shows that paradoxically the more social networks users participate, the more they become "vulnerable" to be controlled and exploited by advertisers and firms.

The ambivalence of participation also comes into sharper focus if its value is considered. It is not automatic that more participation produces better decisions, since in many circumstances an increase in participation might cause a reduction in the quality of policy (slow, poor quality decisions, or incoherent combinations of policies). Perplexities about the value of participation are expressed by many actors also because of the costs that participation carries, but at the same time it is evident that participation brings with it a developmental value in educating people and it enhances "both the meaning of their lives and the value of their social relationships" (McLean, 1966, p.362).

Here, we are interested in participation in its weak meaning. That is we are interested in understanding if those who participate in political, union, religious and volunteering groups use mobile phones in a particular way. We define participation in broader terms as the notion of bridging forms of social capital (Putnam, 2000). As McLean argues, organizations such as unions, religious and volunteering associations "set the context of politics, give their active members administrative experience and are capable of overt political action if their interests or principles are threatened" (1966, p. 362). Putnam (2000, p. 21) also describes a similar idea: "a society characterized by generalized reciprocity is more efficient than a distrustful society.... Trustworthiness lubricates social life." As Fortunati reported (in press), the strong and weak forms of participation are destined to correlate and combine among them with different degrees and modalities. But if the weak forms of participation are scarce and contingent, very rarely will there be a high degree of the strong form of participation. Generally in the mass society a scarce presence of strong participation combines with a high presence of weak participation. The gradual and simultaneous increase of both the forms of participation is considered among the indexes of modernization. On the contrary, the lack of both of them is an indicator of social and political marginality. According to McLean (1966, p. 362), there is evidence that shows that the best places for large participation by people are industrial democracies and the best strategy is the devolution of power to small territorial units.

The broad (and weak) concept of participation is discussed by social scientists that observe the laws and functioning of everyday life. In his first book, the *Theory of Moral Sentiments* (1759), Adam Smith, the father of the modern economics, wrote that that are principles in human nature that make people participate in others' fortune and misery. These engines of social development and civilization are "social passions" such as generosity, empathy and compassion. Even two centuries later, this idea remains current in the works of philosophers such as Martha Nussbaum (2001).

Recent research provides evidence of a general decrease in societal and political participation. In the United States the seminal work of Putnam (2000) documents this decrease through a vast array of different data. In Italy, research published by ISTAT in 2006 shows that political participation, defined as an activity aiming to influence the political choices of a country and the program of political parties and government, is only experienced by very small group of the population (Baldazzi et al., 2006). Similar data attests to the decrease in participation across many industrial countries over the last decades. Nevertheless, two points should be considered. First, social behaviors reflected through participation are dependent on the current national political environments. The rise of social movements, union strategies and even availability and appropriation of ICTs are affected by the government in power. Second, along with objective data, it is important to collect data on people's perception of their social participation as it often differs from actual behaviors. Our research analyses Europeans' perception of their social participation in 2009.

Aim and Method

We researched the diffusion and adoption of popular ICTs among a subset of Europeans who were at the same time investigated as regards social participation in 2009. Data on level and frequency of societal participation, use of the mobile phone and socio-demographic variables were collected by means a phone survey of 7254 participants.

The study, which was conducted in 2009, partially replicated research from 1996 on the five most populous and industrialized European countries, France, Germany, Italy, Spain and the UK (Fortunati, 1998). To simplify our language we will use the term Europe and Europeans to indicate these five countries and their populations. Telecom Italia financed both these studies.

The demographic variables included in the research were gender, age, education, activity, family size and family typology, and the size of the city of residence. Among the respondents, 3471 were males (47.8 percent) and 3783 were females (52.2 percent). The respondents' age was categorized into five groups (14-17, 18-24, 25-44, 45-64 and 65 years and over). Education level was divided into the following categories: low (primary and secondary school diploma), middle (high school diploma) and high (College/University degree or higher). Respondents' occupational classifications included persons who were employed, unemployed, housewives, students and pensioners. The typology of families was divided into singles, couples without children, couples with children, single-parent families and mixed families (all the other types of families). The research found that more than half of the participants belonged to families comprised of one or two people. Finally, four categories were distinguished with respect to city size: cities of 10,000 inhabitants or less, from 10,000 to 100,000 inhabitants, from 100,000 to 500,000 inhabitants and cities with over 500,000 inhabitants.

Results

1. Trends in Participation

The following question was asked in regards to the respondents' perception of their societal participation: "In your spare time, how frequently do you go out to participate in activities promoted by associations and clubs (religious, political, voluntary work and so on)?" This question differs from other questions that distinguish political from religious and voluntary involvement, and was meant to obtain aggregate participatory data. The suggested answer categories were: several times a week, once a week, one or two times a month, less than once a month and no/never. In some analyses the answers were re-classified into two categories only: "yes," which includes "several times a week, once a week, one/two times a month and less than once a month," and "no/ never." Results from the data showed that in 2009 less than half of the European population expressed some form of participation (table 1). Among the participatory respondents, 32.6 percent participated one or two times a month or less. It is important to note that this data only addressed activities conducted outside the home, and does not include the political, religious and volunteer activities carried out inside the home.

Table 17.1 Participation in Political, Social,
and Religious Activities by Europeans in 1996 and 2009

Years	Social Participation		
	Yes	No	Total
1996	2060	4543	6603
	(31.2%)	(68.8%)	(100.0%)
2009	3278	3976	7254
	(45.2%)	(54.8%)	(100.0%)

The data showed that within the democratic countries studied, the majority of the population did not participate socially and that the amount of participation varied significantly among these countries. Italy had traditionally been a very participatory country, credited for having the strongest communist party in Europe, strong unions, a strong tradition of voluntary work and a strong presence of the Catholic Church. However, currently Italy ranks as the country with lowest participation, only 33.6 percent, a drop from 37.4 percent twelve years ago. This consistent decrease was attributed to the deluding performance of the political class, the unions and the public administration as well as to the secularization of everyday life. Italian participation rate is closely followed by that of UK and at a distance of Germany (Table 17.2). Unlike Spain and France, Italy, UK and Germany experienced social participation in less than fifty percent of the respondents. We can hypothesize that Spain and France are more active in empowering democracy through the direct involvement of their citizens.

From a historical perspective there has been a general increase of political, social and religious participation in Europe over the last 12 years (Table 17.1), although this may only represent the respondents' perception not objective reality. If we consider that during these 12 years a broad penetration and appropriation of ICTs occurred within the studied countries, we can hypothesize that social actors have taken advantage of these technologies, which facilitated their participation.

When analyzing each country separately, we find that the change has not been homogeneous and varies over time. While in 1996 the more active countries were Italy and Germany, twelve years later, Italy is the only country that shows a decrease in participation (Table 17.2). While Italy has the lowest percentage of participation, in France it more than doubled, and in Spain it increased by a fifth.

Table 17.2 Participation in Political, Social, and Religious
Activities in the Five European Countries 1996 and 2009

Countries	Social Participation			
	1996		2009	
	yes	no	yes	no
Italy	515	861	470	928
	(37.4%)	(62.5%)	(33.6%)	(66.4%)
	(25.0%)	(19.0%)	(14.3%)	(23.3%)
Germany	632	1134	895	1024
	(35.8%)	(64.2%)	(46.6%)	(53.4%)
	(30.7%)	(25.0%)	(27.3%)	(25.8%)
France	321	1012	810	613
	(24.1%)	(75.9%)	(56.9%)	(43.1%)
	(15.6%)	(22.3%)	(24.7%)	(15.4%)
UK	302	879	528	883
	(25.6%)	(74.4%)	(37.4%)	(62.6%)
	(14.7%)	(19.3%)	(16.1%)	(22.2%)

Table 17.2 (*continued*)

| Countries | Social Participation | | | |
| | 1996 | | 2009 | |
	yes	no	yes	no
Spain	290	657	575	528
	(30.6%)	(69.4%)	(52.1%)	(47.9%)
	(14.1%)	(14.5%)	(17.5%)	(13.3%)
Total	2060	4543	3278	3976

The first line of percentages adds up by row and by year to 100% each and the second line adds up by column to 100%.

$\chi^2 = 100.50$, df $= 4$, p < 0.0001 for 1996.

$\chi^2 = 212.109$, df $= 4$, p < 0.0001 for 2009.

2. The Profile of Participatory Actors

We define *participatory actors* as respondents who have participated in social organizations. From the sample of 7254 respondents, 3277 (45.2 percent) are participatory actors. In order to understand the influence of mobile technologies on the lifestyle of the population, we investigated the demographic and sociological profile of participatory actors. As hypothesized, results showed that males participate more frequently than females[1] and that the higher the education level of the respondent, the greater their participation.[2] We conclude that education positively affected the level of social involvement. City size and location also had an effect on participation: the more populous the city, the more widespread the participation.[3] Contrary to many qualitative research projects that indicate that the core of political and social activity is the community, our results showed that urbanity plays a more important role. A certain mass of people is required to effectively develop "social passions" such as compassion, empathy and generosity which are strong motivators for participation. This suggests that society should reflect more on the effect of urbanity on participation, in particular in Italy with the emergence of new territorial parties, such as the *Lega Nord per l'Indipendenza della Padania*.

The effect of respondents' age on participation varied.[4] For the youth involvement might be associated with their enthusiasm, sense of justice and availability of time, while for the older age respondents, sensitivity to social commitments might be brought about by maturity and wisdom.

Occupation had an expected effect on participation. The employed exercised the highest participation, followed by students.[5] The unemployed, the retired and in particular the housewives were the most reluctant participatory actors. The typology and size of respondents' family did not have a significant correlation to social involvement.[6]

To properly sketch the profile of the European participatory actor at a sociological level, we also needed to understand how personal time and activities were affected by social participation.[7] The data showed that participatory actors were moderate in the time allocated to their personal activities, while those who did not participate socially were distributed at different poles. The former said they visited friends at most once a week;[8] the latter either said they never saw friends or saw them several times a week.[9] The data also showed that non-participants were more likely to go out with friends, play sports and read newspapers daily, whereas participatory actors did less of these activities. However, while participatory actors did not reduce the variety of forms of communication practices, they did use less the various forms that they had available.

3. Participation and the Role of Information and Communication Technologies

In this section we explored whether social participation is associated with a specific type of ICT by cross-referencing social participation with a preference for particular communication media. The following communication technologies were suggested: landline (fixed) telephone, mobile (cellular) telephone, radio, television, personal computer and Internet. Internet was ranked first, followed by television, landline, mobile phone, and finally the computer and radio (Table 17.3).

Table 17.3 Social Participation and Media Preference

| Media preference | Participation | | |
	Yes	No	Total
Internet	1033	1388	2421
	(42.7%)	(57.3%)	(100.0%)
	(31.5%)	(34.9%)	(33.4%)
Television	633	845	1478
	(42.8%)	(57.2%)	(100.0%)
	(19.3%)	(21.3%)	(20,4%)
Landline	564	606	1170
	(48.2%)	(51.8%)	(100.0%)
	(17.2%)	(15.2%)	(16.1%)

Table 17.3 (*continued*)

| Media preference | Participation | | |
	Yes	No	Total
Mobile	422	417	839
	(50.3%)	(49.7%)	(100.0%)
	(12.9%)	(10.5%)	(11.6%)
Computer	302	373	675
	(44.7%)	(55.3%)	(100.0%)
	(9.2%)	(9.4%)	(9.3%)
Radio	322	347	669
	(48.1%)	(51.9%)	(100.0%)
	(9.8%)	(8.7%)	(9.2%)
Total	3276	3976	7252

The first line of percentages adds up by row and by year to 100%. The second line adds up by column to 100%.

It is interesting to note that although the mobile phone industry has experienced undeniable success, the traditional landline telephone remains the preferred method of communication among the respondents. In addition, with the advent of the Internet, media preferences have been restructured. In the survey conducted in 1996, although only four devices were listed (television, radio, telephone and computer), the ranking was quite different from the current results: television ranked first with 42.8 percent, followed by radio (26.7 percent), telephone (20.9 percent) and computer (9.7 percent) (Fortunati, Manganelli, 1998).

These results showed that those who are involved in some form of social participation preferred the telephone (both landline and mobile) and the radio when compared to those who were not participatory actors, while the latter preferred television and the Internet (Table 17.3). We concluded that social participation is mainly associated with oral communication and the supporting technologies, while non-participants associated better with visual, unidirectional communication.

We also analyzed how the respondents associated ICTs with their motive of usage, namely for amusement, for facilitation of social relationships or for help with time management. (These three purposes came out from a pretest we did before the first survey in 1996.) The results were unexpected. Table 17.4 shows how participatory actors and non-participants ranked media on a 5-point scale (where 1= not at all important and 5 = very important).

Table 17.4 Average Scores for the Three Main Purposes of ICTs Use: Amusement, Facilitation of Social Relationship, and Help with Time Management of Participatory Actors and Non-Participatory Actors

	Participa- tory Actors	Non- participa- tory Actors	Overall means	t	df	p <
Facilitation of social relationships						
Telephone	3.75	3.63	3.69	4.37	7252	0.0001
Mobile	3.59	3.22	3.38	11.64	7252	0.0001
TV	3.13	3.17	3.15	−1.29	7252	n.s.
Internet	3.26	2.89	3.06	10.37	7252	0.0001
Radio	2.99	2.86	2.92	4.41	7252	0.0001
Computer	3.02	2.72	2.86	9.00	7252	0.0001
Amusement						
TV	3.60	3.77	3.69	−6.36	7252	0.0001
Telephone	3.56	3.43	3.50	3.27	7252	0.001
Radio	3.31	3.23	3.26	2.49	7252	0.05
Mobile	3.29	2.95	3.10	10.73	7252	0.0001
Internet	3.15	2.81	2.96	9.31	7252	0.0001
Computer	3.07	2.75	2.90	9.52	7252	0.0001
Help in time management						
Telephone	3.46	3.37	3.41	2.91	7252	0.01
Mobile	3.42	3.14	3.27	8.46	7252	0.0001
Internet	3.10	2.77	2.92	9.40	7252	0.0001
TV	2.89	2.91	2.90	−.76	7252	n.s.
Computer	2.95	2.62	2.77	10.12	7252	0.0001
Radio	2.79	2.66	2.72	4.00	7252	0.0001

All three motives averaged around 3 on the scale: amusement (M = 3.03), facilitation of social relationships (M = 3.02) and help with time management (M = 3.00). The results showed that at the time of the survey (2009) the listed ICTs were moderately relevant to achieve these motives. Compared to 1996, television, radio and landline telephone preference had decreased while the mobile phone had increased. The core technologies for the facilitation of social relationships and for helping with time management were the landline telephone and the mobile phone, while television and telephone were the preferred means for amusement. The results of the "t test" for independent samples showed that the average scores were higher for participatory actors than for non-participants, with the exception of the television, which was the main source of amusement for those who do not socially participate.

4. The Emotional Connection to the Mobile Phone

We also addressed the following questions:
- Is social participation associated in some way to a certain emotion towards the mobile phone?
- Do participatory actors differ from the non-participatory ones with respect to the amount and type of ICT used?
- And if so, what are the perceived effects of the use of mobile phones on the social environment?

In regards to the first question, our results showed that participatory actors feel more enthusiasm, pleasure, interest and curiosity for mobile phone than non-participants. In contrast non-participants used the mobile phone more as a tool for keeping in touch.[10] It appears that the more limited the social sphere, the more the phone is used as a substitute for social relationships.

Our data showed that people who actively participate made a higher number of outgoing and incoming mobile calls and sent a higher number of SMS messages compared to non-participants.[11] We can conclude that social participation is associated with a more intense use of the mobile phone.

However, when we checked the typology of people participants called, we found that only 3.9 percent used the mobile phone to call other members of associations etc. This percent was even lower for non-participatory actors: 1.7 percent. Therefore, the intense use of the mobile phone by participatory actors has a more general purpose rather than being connected to the specific activity of participation. Participatory actors called association members, colleagues and friends outside their family circle more frequently, while non-participants called family more often.

We now consider the main mobile phone functions: SMS, Multimedia Messaging Service (MMS), where messages are sent with attached photo/video/audio files, email, chatting, looking through the news (receiving SMS or MMS with news), taking photos, making videos, listening to MP3s, playing games, voice recording, Bluetooth, radio broadcasting, instant messaging, Internet browsing, downloading logos and ringtones, downloading software, using the handset as a modem, clock, text editing (writing with T9), alarm clock.

Our index explores the low, medium and high level of usage of the mobile phone functions by our respondents. The results showed that participatory actors were more sophisticated in the use of mobile phone services,[12] with a mean of 7.53 functions compared to a mean of 6.24 functions used by those who do not socially participate.

Finally, we researched how respondents perceived the effects of mobile phone use on the social lives; we posed the following questions:

Q1. From your personal experience, can you tell me if the use of the mobile phone encourages or reduces face-to-face communication between you and your acquaintances?

Q2. From your personal experience, does the use of the mobile phone render the communication more profound or more superficial?

Q3. From your personal experience does the use of the mobile phone increase or decrease your social circle?

These questions deal with an important issue in the scientific community debating the way in which the use of the mobile phone has re-addressed social relationships. Here we were particularly interested to know how this re-addressement is evaluated by both participants and non-participants.

While half of the respondents stated no change in their social relationships, we found particularly interesting that participatory actors believed that the mobile phone had increased their face-to-face encounters[13] (Table 17.5). The perception is that the use of the mobile phone is a player in building social relationships.

**Table 17.5 Social Participation and
the Effect on Body-to-Body Encounters**

	Participation		
Face-to-face encounters	**Yes**	**No**	**Total**
We meet more	887`	905	1792
	(49.5%)	(50.5%)	(100.0%)
	(32.3%)	(27.5%)	(29.7%)
We meet less	731	838	1569
	(46.6%)	(53.4%)	(100.0%)
	(26.6%)	(25.5%)	(26.0%)
Interactions are the same	1125	1548	2673
	(42.1%)	(57.9%)	(100.0%)
	(41.0%)	(47.0%)	(44.3%)
Total	2743	3291	6034

The first line of percentages adds up by row and by year to 100%. The second line adds up by column to 100%.

In response to the second question, data showed that there is no significant difference in opinion between participants and non-participants regarding the quality of mobile phone conversation. Over one third of respondents stated that the quality of mobile phone communication remained unchanged, nearly one third said that it has become more superficial and little more than one quarter believed it had become more in depth. This shows there is a pessimistic view concerning how the quality of conversation in the mobile age has developed. The outcome was not surprising as mobile communication has always been judged as less communicative than body-to-body conversations (Fortunati, 2002b).

The answers to the third question produced similar results showing that 45 percent of respondents believed the mobile phone had not affected their relationships and 43.6 percent were convinced that their social circle was increased. Only 11 percent of respondents, of which the majority were active participants, felt that mobile phone use had decreased their social circle[14] (Table 17.6).

Table 17.6 Social Participation and Its Effect on the Social Circle

Social sphere	Participation		
	Yes	No	Total
Enlargement	1192	1439	2631
	(45.3%)	(54.7%)	(100.0%)
	(43.4%)	(43.7%)	(43.6%)
Reduction	362	301	663
	(54.6%)	(45.4%)	(100.0%)
	(13.2%)	(9.1%)	(11.0%)
Social sphere is the same	1190	1552	2742
	(43.4%)	(56.6%)	(100.0%)
	(43.4%)	(47.1%)	(45.0%)
Total	2744	3292	6036

The first line of percentages adds up by row and by year to 100%. The second line adds up by column to 100%.

We conducted a series of multiple regression analyses with the purpose of verifying if social participation influences the use of the mobile phone when gender, age, education and size of the city of residence, were the controlled variables. The dependent variables were: the number of mobile calls, SMSs and the set of mobile phone functions used by interviewees.

The results indicated that the frequency of participation in political, social and religious activities was a variable that positively influenced the use of mobile phones, in particular for voice calls (for outgoing calls, $\beta = .061$, $p < 0.0001$; for incoming calls, $\beta = .054$, $p < 0.0001$) and for the number of mobile phone functions used ($\beta = .090$, $p < 0.0001$).

However, mobile phone ownership was not related to social participation. The profile of the 16.8 percent of respondents that did not to possess a mobile phone was predominantly females, those aged 45 or more, with low education level, pensioners, singles, and people living in cities with less than 100,000 inhabitants.[15]

Final Discussion and Conclusions

This research showed that during the past 12 years there has been a general increase of social participation in the five countries surveyed in Europe. European society in the whole seems to be more convincing than in the past in involving citizens in social participation. This probably means that current participation is more able to develop its positive features. However, in 2009 less than half of Europeans participated socially and this was dependent on cultural characteristics and time frame. Despite its increase, social participation remains an experience carried out by a minority of population. The demographic profile of participatory actors was predominantly made up of men, those with higher education, those living in highly populated cities, and either students or the employed. Only the most advantaged people continue to be candidate for social participation.

Participatory actors decreased the amount of time for family and other social interactions in order to dedicate more time to social involvement. They preferred oral communication and technology that supports social relationships, whereas non-actors leaned towards visual communication and technology with unidirectional messages. This connection between voice and social participation deserves further investigation and reflection.

The three main purposes identified for ICT use, facilitation of social relationships, amusement and help with time management, consistently ranked higher among participatory actors, with the exception of television, which was the primary source of entertainment for non-participants. While mobile phone use had increased since 1996 and landline telephone use had decreased they remained the core technologies for the facilitation of social relationships and for time management.

Our research found that participatory actors showed more enthusiasm, pleasure, interest and curiosity towards the mobile phone. People who actively participated made a higher number of mobile calls and sent a higher number of SMSs compared to those that are not socially active, but this more intensive use was not related to performing participatory activities. Participatory actors use the mobile phone to build relationships outside their social circle as well as to maintain relationships with friends and family members who they see less often. Participants were found to be more sophisticated in their use of mobile phone functions and were convinced that the use of the mobile phone had brought an increase in their face-to-face interactions. However they have a pessimistic view of how the quality of conversation has developed in the mobile age. Finally, the frequency of participation in political, social and religious activities positively influences the use of the mobile phone, also attracts a more sophisticated use of its functions and shapes in a particular way attitudes and emotion towards this device.

In the same period in Europe a broad penetration and appropriation of ICTs occurred, which lead us to conclude that there is a close relationship between the appropriation of these technologies and an increase in social participation. But this relationship is not direct, but oblique. In fact this research shows that participatory actors need to use the mobile phone more intensively and sophistically than non-participants for organizing their everyday life that is evidently more complex. The use of the mobile phone is functional to the increase in the number of participatory actors but the mobile phone serves to a limited extent to support their participant activities. Social participation in its strong and weak meaning has still to discover how to make use of the mobile phone.

All kinds of social policies that aim to foster citizens' participation have to face the "problematic side" of social participation, from which people receive often frustration and delusion, and have to invent a participatory use of the mobile phone.

Notes

1    $\chi^2 = 26.976$, df $= 1$, $p < 0.0001$.
2    $\chi^2 = 217.468$, df $= 3$, $p < 0.0001$.
3    $\chi^2 = 206.286$, df $= 4$, $p < 0.0001$.
4    $\chi^2 = 43.132$, df $= 4$, $p < 0.0001$.
5    $\chi^2 = 117.454$, df $= 5$, $p < 0.0001$.
6    $\chi^2 = 45.178$, df $= 5$, $p < 0.0001$ and $\chi^2 = 32.423$, df $= 4$, $p < 0.0001$.

7     $\chi^2 = 45.178$, df = 5, $p < 0.0001$ and $\chi^2 = 32.423$, df = 4, $p < 0.0001$.
8     $\chi^2 = 186.490$, df = 5, $p < 0.0001$.
9     $\chi^2 = 174.729$, df = 5, $p < 0.0001$.
10    $\chi^2 = 286.647$, df = 28, $p < 0.0001$.
11   For the outgoing mobile calls: t = 5.82, df 5470, $p < 0.0001$; incoming mobile
      calls: t = 5.65, df 5396, $p < 0.0001$; for outgoing SMS: t = 2.56, df 3731, p <
      0.05; for incoming SMS: t = 2.02, df 3837, $p < 0.05$.
12   t = 10.52, df 6033, $p < 0.0001$.
13    $\chi^2 = 24.854$, df 2, $p < 0.0001$.
14    $\chi^2 = 27.063$, df = 2, $p < 0.0001$.
15   For gender: $\chi^2 = 37.680$, df = 1, $p < 0.0001$; age: $\chi^2 = 522.648$, df = 4, $p < 0.0001$;
      education: $\chi^2 = 48.428$, df = 3, $p < 0.0001$; activity: $\chi^2 = 604.565$, df = 5, $p < 0.0001$;
      family typology: $\chi^2 = 94.818$, df = 5, $p < 0.0001$; size of the centre of residence:
      $\chi^2 = 34.268$, df = 4, $p < 0.0001$.

References

Castells, Manuel, Qiu, Jack, Fernandez-Ardevol, Mireia, & Sey, Araba (2006) *Mobile communication and society. A global perspective.* Cambridge, MA: MIT.

Castells, Manuel (2007) Communication, Power and Counter-power in the Network Society, *International Journal of Communication*, 1, pp. 238-266.

Baldazzi, Barbara, Montecolle, Silvia, Orsini, Sante, Rivellini, Giulia, Savioli, Miria, Scalisi Pietro (2006) Partecipazione politica e astensionismo secondo un approccio di genere. Accessed 15 February 2009 at: www.istat.it/istat/eventi/2006/partecipazione_politica_2006/partecipazionepolitica.pdf.

Farinosi, Manuela (2010) Verso una prospettiva post-panottica. Privacy, controllo e contenuti generati dagli utenti (UGC). Tesi di dottorato, University of Udine.

Fortunati, Leopoldina (2002a) "The mobile phone: towards new categories and social relations," *Information, Communication & Society,* vol. 5, 4, pp. 513-528.

Fortunati, Leopoldina (2002b) Italy. Stereotypes: true or false, in James Katz, Mark Aakhus (eds) *Perpetual Contact: Mobile Communication, Private Talk, Public Performance.* Cambridge University Press, Cambridge, pp. 42-62.

Fortunati, Leopoldina (2006) User Design and the Democratization of the Mobile Phone, *First Monday,* 7(9).

Fortunati, Leopoldina (2007) Immaterial Labor and its Machinization, *Ephemera. Theory & Politics in Organization,* 7(1), pp. 139-157.

Fortunati, Leopoldina (ed.) (1998) *Telecomunicando in Europa.* Milano: Angeli.

Fortunati, Leopoldina, Manganelli, Anna Maria, (1998) La comunicazione tecnologica: Comportamenti, opinioni ed emozioni degli Europei, in Fortunati, Leopoldina, (ed.), *Telecomunicando in Europa,* Milano: Angeli.

Fortunati, Leopoldina (forthcoming) Online Participation and the New Media. In Hajo Greif, Larissa Hjorth, Amparo Lasen, Claire Lobet-Maris (eds.) *Cultures of Participation.* Berlin: Peter Lang.

Gallino, Luciano (1993) *Dizionario di Sociologia.* Milano: Tea.

Katz, James E., Aakus, Mark (eds.) (2002) *Perpetual Contact: Mobile Communication, Private Talk, Public Performance.* Cambridge: Cambridge University Press.

Le Bouedec, George (1984) Contribution à la méthodologie d'étude des représentations sociales, *Cahiers de Psychologie Cognitive,* vol. 4, 3, pp. 245-272.

Ling, Richard, Donner, Jonathan (2009) *Mobile Communication.* Cambridge: Polity Press.

McLean, Ian (1966) *Concise Dictionary of Politics.* Oxford: Oxford University Press.

Nussbaum, Martha (2001) *Upheavals of Thought. The Intelligence of Emotions*. Cambridge: Cambridge University Press.

Nyíri, Kristof (ed.) (2003) *Mobile Democracy. Essays on Society, Self and Politics*. Vienna: Passagen Verlag.

Putnam, Robert D. (2000) *Bowling Alone. The Collapse and Revival of American Community*. New York: Simon & Schuster.

Rheingold, Howard (2002). *Smart Mobs. The Next Social Revolution*. Cambridge: Perseus Books.

Sarrica, Mauro, Grimaldi, Floriana, Nencini, Alessio (2009) Rappresentazioni Sociali della Partecipazione. Una ricerca esplorativa con giovani veneziani. In: International Conference 'Giovani e Società', Forlì, 26-28 March.

Scott, John, Marshall, Gordon (2005) *Dictionary of Sociology*. Oxford: Oxford University Press.

Smith, Adam (1759) *Theory of Moral Sentiments*.

Vincent, Jane, Harris, Lisa (2008) Effective Use of Mobile Communications in e-Government: How do we reach the tipping point? *Information, Communication & Society*, 11(3), pp. 395-413.

Wellman, Barry, Hogan, Bernie (2006) Connected Lives. The Project. In Patrick Purcell (ed.) *Networked Neighbourhoods*. London: Springer.

18

Technological Rabbits
and Communication Turtles

Irving Louis Horowitz

In the first class of my first course that I took in philosophy at the City College of the City University in New York, the instructor, a fine scholar by the name of Mortimer Kadish, who went on to a distinguished career at Case Western Reserve University, asked a seemingly harmless question: "why do we look upon the mathematics of Archimedes as a curious artifact, and the ethics of Aristotle as an essential text in the learning process of our age?" I recollect that no one in that class was able to understand the epistemological significance, much less answer that question! I am not certain that I can resolve this paradox even at this point in time, but at least I now know the burden of such a question.

What prompts such a stark recollection to a dim past is a review of a publication entitled *Technology First: Journal IV* with the inevitable theme of "The Next Step in Communication." The issue features articles on consumer demand for cheaper, easier broadband access, with Wi-max or the tongue twister: worldwide interoperability for microwave access. This is followed by a piece on the greater use of flame retardants to reduce fires and injuries that come from electrical faults. It emphasizes the use of HIPS or high impact polystyrene and Acetonitriler Butadiiene Styrene. This is followed by information on the latest devices to test functionality and connectivity of the entire UBS protocol on a single chip. A follow-up essay discusses how, in an age in which 90 percent of people in the United Kingdom have a cell phone and 50 percent have Internet access, the issues sharply change from communication between people, with devices, and even how devices communicate with each other, to the growth of wireless connectivity as an end unto itself—thus replacing cables and the need to increase security.

The pattern of such empirical research is now well established and doubtless known to most of the people working in the field of information technology and communication. There is little point in reciting all of the amazing developments taking place that truly are changing the shape of the physical and human environment. It is evident that technical applications are the critical factor reshaping the character of our lives: from how we read, what we feel, and when we perform certain routine and extraordinary actions. Entire industries are dissolving before our eyes: from newspaper closings to automobile collapses. At the same time, and in reverse entire industries are coming into existence: from open access to information in a variety of electronic forms to automobiles driven by miniaturized battery devices that beckon to make petroleum an obsolete fuel. Everything from the nature of education to the power of extending human life through medical devices and drug regimens is at stake in such changes.

In the decision-making process, communications experts and social scientists provide different qualitative inputs. The former provides expertise that tends toward non-ideological, non-political solutions to social problems. Systems design displace ideological disposition for many in the field. Social scientists, for their part, tend toward political solutions. The latter are usually wary of technological bureaucracies sponsored by governments as technicians are of charismatic or quasi-patriarchal domination by governments.

This divide between communication research and social scientific studies was not always the case. For many years, especially in the middle of the last century, communication studies were very much a part of sociology and to a lesser extent political science. People like George Gerbner, Ithiel de Sola Pool, Elie Abel, Hans Speier, Harold D. Lasswell, and Hugh Dalziel Duncan, to name a few, very much participated in the sociological imagination. As the field of communication studies advanced its own academic and research agenda, it developed strong empirical tendencies, derived in some measures from public opinion survey research and propaganda analysis across boundaries. Such a movement toward the quantitative was stimulated if not entirely supported by giant corporations in the field of communications. In short, a field that hardly existed at the turn of the twentieth century took on special service characteristics that moved it away from social science as a source of intellectual energy.

Connections to older social sciences were seriously weakened over time. The examination of moral issues has just about vanished among no-nonsense technicians. Communication studies became a field unto itself, with strong statistical and quantitative guidelines that set strong boundaries to what was considered viable or valuable. These parameters became the measure of professionalism, and one would have to say that in numerical terms this great divide allowed for the growth of communication studies to a point that it has dwarfed those areas from which it emanated. This is hardly an unprecedented phenomenon.

With success come new problems. Moving away from one's own history is a mixed blessing. The problem can be stated as follows: the communication impulse is toward rational, systematic solutions, and toward impatience with the norms and values of non-rational and ideological, class, radical, and ethnic interests. Like most professional groups, people doing research on communication have been reticent to perceive themselves as yet another special interest group whose claims to political preeminence are neither more nor less valid than the claims of other interest groups. The communication field sees itself as largely a professional group, part of the great expansion of technocracy no less than technology. Indeed, with a glacial shift from the means of production to the means of communications, and specifically from measuring wealth and power in terms of proprietary rights rather than inventory of products, the impulse toward technocracy becomes sharper, and assuredly more transparent.

The most vital issue immediately affecting technologists and social scientists alike is their joint commitment to developmental application, to framing a language amid a set of techniques widely understandable because they rest on results based upon commonly employed systems and designs. Thus far, the scientific language of system design and man-power allocation has not been fully integrated with a political language of social transformation and decision-making. This is not to fault either professional group. Indeed, social scientists have reinforced present methodological divisions by assuming stylistic differences between systems management and applied research to be permanent simply because they are operational. The emergence of computer technology, widely adapted to social science uses, and the corresponding emphasis on measurable indicators of health, welfare, fertility, crime, etc. has clearly linked the two groups so as to make new breakthroughs and cooperation highly probable in the next stage of communication studies. But these breakthroughs are linked to practical applications, such as security concerns and privacy issues in law and behavior.

Any serious dialogue between social scientists and communication experts requires a recognition of the enormous contributions made by modern technology to social science such as: (a) the engineering emphasis on the special problems involved in applications and direct research, which has created a whole new field of social planning; (b) the managerial emphasis on posing problems in soluble formulations, rejecting apocalyptic readings of events as a way of bringing about a solution to problems of the world; (c) the emphasis on design, delivery and organization, particularly with respect to the processes of urban expansion and industrialization. This in turn has permitted social scientists to place greater emphasis on anticipating problems of social reorganization by conceiving the larger society as appropriate for the input of parallel design, delivery and organization. The devil in this bargain is the loss of interest in larger ethical domains—both for mass communication and social research.

The reference to a link between humanities issues and the social sciences has itself come to be viewed as an indicator of parochialism. Technicians often think of social studies apart from an understanding of social sciences. Further they tend to view social science inputs as shadow rather than substance. For example, many important writers see social problems strictly in terms that can be treated by technicians, i.e., environmental pollution, transportation, water resources, medical care, regional development, etc. No mention is made of problems of revolution, terrorism, anomie, generational conflict, racial struggle, etc. The view of the communications personnel remains insular and cautious to a fault. It is clear that as long as commonplace pieties are used to describe the social sciences, the contributions of the history of social science to communications research, with specific reference to developmental analysis, will remain peripheral, largely unexamined, and perhaps an irritant.

There is a strong tendency on the part of information technicians to celebrate their intellectual self-sufficiency. They now speak of a doubling of knowledge in decades. They refer with pride to the fact that a history of technology covering two earlier centuries may take one volume, while the same sort of historical survey of the twentieth century alone may be a multi-volume affair. They even manage to get some scholars to promote the heady wine of futurology, or "one hundred technical innovations likely in the next thirty-three years." The more judicious engineers speak in sober and cautious terms, pointing to the relative slowdown in new inventions in recent years, and the under-utilization of present plant capacities. The less cautious among them speak of how robot technology displaces human will in the execution of the tasks of armed conflict and open warfare.

What has taken place, and what obscures the gulf between disciplines is the sophisticated refinement of established inventions; or more precisely a lag between invention and applications. In the first half of the twentieth century major inventions were realized and widely applied: plastics, automobiles, radios, television, and commercial aircraft. In the second half one may list electronic computers, wireless transmissions, videotape, nuclear explosives, robotics, and antibiotics. We may now be entering a period of invention refinements rather than breakthroughs. If there has been a slowdown, as a sober reckoning would in fact indicate, and if we are getting a phase of application in place of innovation, then it is important that social dimensions of the problem of innovation be raised and celebrations be modified.

The shift in emphasis from the means of production to the instruments of communication is highlighted by the fact that communication as such changes the character, design and structure of production as such. The development of wireless communication illustrates this shift. In the airline industry passengers can now use cell phones as two-dimensional bar codes that can be displayed and scanned at boarding gates and security checkpoints. In the auto industry, all phones can now start the car and adjust the seats; digital mobile keys are in the making. Bigger screens and better browsers are now allowing smart phone users to do online banking. In education, the mobile phone is equivalent to the laptop, and hence an instrument for basic changes in pedagogic techniques. In energy, the incorporation of smart meters now permits homeowners to control energy usage and consumption. Medical records are increasingly incorporating mobile devices to maintain records, adjust drug usages, and text message patients directly. Mobile Web bookings have become commonplace in the hotel industry. Political campaigns increasingly rely upon text messaging with supporters and to recruit volunteers in mobilizing for critical campaigns. Finally, closer to home, publishing has witnessed the rise of a new era of digital, mobile users that permit wireless data modems that change procedures for everything from pricing production to reading classics.

An impressive fact is that new inventions and applications are increasing at so great a speed that they cannot be readily absorbed at quotidian human levels. More often than not, they require collective efforts that far transcend the capacities of any single person, or sometimes even an industry. No less significantly, the sort of inventions now being brought out cannot be counted on to always produce rapid returns on financial investments. Hence, private firms are reluctant to invest in large-scale research, preferring refinements of presently marketed items. Public sector enterprises, for their part, are hamstrung by the lack of funds, or by constant pressures within certain societies against extended investments in new projects that may have little chance of commercial success. Because of this gap between economy and technology, at least in the area of commodities, the gap between invention and wide scale usage must be viewed as structural and long-range in character, rather than representing a temporary lag in the overall economy.

Instead of coming to terms with such essential socio-structural issues, what Robert Boguslaw termed the new utopians—those technicians turned social analysts—have developed an exaggerated burst of optimistic appraisals about the social structure. The shift in technocratic orientations is away from the social role of scientific specialists, toward a belief that political or administrative responsibility may become a relatively minor prize in the environment game. In a new view of life the technological vision of social organization has gone along with utopianism. The task of the human engineering principle will be in matching the constructed environment to the person rather than vice versa.

To date, this relation of the person to the machine has been a mismatch. An intriguing piece on "Faux Friendship" by William Deresiewicz, gets to the heart of the dilemma. "Actual human contact rendered 'unusual' and weighed by the values of a systems engineer. We have given our hearts to machines, and now we are turning into machines. The face of friendship in the new century." This claim turns the French Enlightenment theorists who argued for "man as a machine" on their heads: "Humanism"—that much maligned word, is now seen by some as a struggle *against* the machine. At one level, the problem resides in the metaphor. Yes, human beings are machines. Then again, they are also animals. But it is clear the whole is far greater than these parts, or if you will, metaphors.

Behind this variety of reductionistic scientific vision are long standing Platonist echoes of the military and intellectual roads to power. Bluntly, we are never told how those doing the "matching" of technology and morality will be chosen. In keeping with this utopian republic underscored by mechanical Power Point displays is a downgrading of entrepreneurial or human administrative skill. Multiple computer uses, and now wireless electronics, make it commonplace to shift the source of error to machines rather than to men and women. In the brave new technological complex, leadership will be done away with, since in a self-sufficient, self-confident, highly educated democracy, all one wants are results. The continued emphasis within the social sciences to ideological explanations is not only dangerous unto itself, but postpones the inevitable coming together of technological and sociological levels of analysis as a means to some larger ethical purposes.

These various developments have led to a new definition of literacy; or computer literacy. This in turn has led to a huge gap between haves and have-nots in the way of modern society as such. Entire segments of society, often those in poorer, working class, and minority situations, are not able to understand, much less utilize the revolution in communications. As a result, any notion of egalitarianism gives way to new forms of social divisions, often with potent consequences. Platonic echoes become pandemic, an acceptable way of distinguishing the gold people from the bronze people. It is risky to speak of education as the mechanism for overcoming such distinctions, especially in a world in which not every person sees education as a panacea to a proper way of life. Indeed, the issue of what constitutes a free society itself is subject to re-examination in the light of new developments. At present, the concept of freedom has been dangerously reduced to information processing.

This brings us full circle to the nature of this contribution to ethical concerns. In the midst of such volcanic change, what is permanent? What is durable? What is even worth preserving? What is the legacy bequeathed to the present? In a world that has elected presidents on the theme of change, what do we do with the permanent things? Indeed, are there such things, or is simply a clinging to the past, a vain expectation that our inheritance itself is something more than an artifact? Whatever the answer is, it is clear that it will not be found exclusively in technology first, last or always! It is my hope that at least some sensitivity to such matters can be at least a small part of the rush to the future.

In an essay on "Engineering and Sociology" some forty years ago, I expressed my debt to Max Weber by differentiating the technological from the social, the mechanical from the human. But it was then an unresolved framework. Only belatedly did I discover the great work of the autodidact Lewis Mumford, and of the geographer Jean Gottmann. It was these outsiders who added red meat to the dry bones of communication theorizing. All of these figures share a common debt to a more distant past. In considerable measure the problem of the disjunction between the technical and the moral resides in an underestimation of that great figure in the history of philosophy, Immanuel Kant. He was the primary source of both Weber's thinking and that of Hannah Arendt. He uniquely appreciated the fact that while change is part and parcel of thought, the rates and styles of such thinking are by no means moving in lock step. This brief set of remarks is a means to correct that oversight.

From Aristotle to Kant we learn of the dangers of dogmatism, of a vision of the scientific and religious orders that proceed without taking into account the sturdiness of quotidian ethical guidelines. The weakness of my earlier formulations stemmed from trying to assign to sociology a level of specific that could correct the abstract nature of technological change. The social is not yet the normative. That is reserved for the Kantian transcendental a priori, the nominal world that is the ghost in the machine of the phenomenal world in which communication travels.

There is a second line that runs from Kant to Arendt. In her remarkable trio of studies on the *Life of the Mind*, she reminds us that in the faculty of judging, as Kant indicated, there is a gap between thinking and judging, between consciousness and conscience. I believe it is fair to say that this extends to the difference between technology and philosophy. This is not about fatuous dualisms, or unmanageable gulfs, but a recognition that all specific fields of endeavor answer to ethical judgment. Arendt summarizes the point well. "The manifestation of the wind of thought is not knowledge, it is the ability to tell right from wrong, beautiful from ugly. And this, at the rare moment when the stakes are on the table, may indeed prevent catastrophes, at least for the self." Such a view does not destroy the grounds for discovery and innovation, but it does place such technical feats in a broader context of the history of human thought.

The grounds for a renewed integration of diverse fields stems from a recognition that although the permanent is the dialectical opposite of the transient, or more bluntly, the eternal is the converse of the changing, both have a place in the normative scheme of things. It is well enough to speak of the inevitability of change—either in terms of improvement or deterioration. And while that is certainly the case, the inverse of that commonplace, and the eternality of permanence is more difficult to accept. The notion of permanence involves a metaphysical presumption of durability, things or ideas beyond those that are subject to quantitative or empirical improvement in demonstrable doses. The strength of the Kantian categories derives not from assumptions about teleological judgment or transcendental mandates, but rather it is based in the long tradition of mathematical suppositions about the axiomatic nature of numbers, and the admittedly difficult assumption that these can somehow be translated into useful moral categories of right and wrong, good and evil, truth and beauty. We can operationally divide these, dismiss some of them, but they have a peculiar ability to pop up in the strangest places.

The cases of abortion and stem cell research are good indicators of how ethical norms change over time. The former is accepted as a contraceptive device of last resort, the latter is understood to be highly risky but a necessary genetic alteration that possibly may save lives. But such procedures are now widely accepted nonetheless. So too are the multiple uses of the cellular phone: to keep parents and their offspring in touch with each other, and also to permit the worst criminal and murderous elements in society from avoiding being captured! The political realm in advanced nations does not celebrate several million cases of abortion annually nor does society adopt mass eugenics as a strategy, but it dares not use risk factors in innovation to deny the need for legal safeguards to unwed mothers, innocent citizens, or the restorative capacity to the damaged limbs of children.

The point of this exercise is to note that moral order does exist and changes exist in its structure. Clearly, rooted values do not change with the volcanic rapidity of developments in the technological order. So we are faced not so much with an oppositional framework of the technical and the ethical, but rather measuring different rates of change, different speeds at which change take place. This is an important lesson in democracy, since it allows for civility in interaction, and liberality in framing rules for exchanges between systems as well as individuals. The great past wars of science versus religion and theology stem from a belief that they are diametrically at odds—with the assumption that science represents freedom and religion dogmatism. But a close look at the utilitarian nature of computers would doubtless indicate that they are as widely used by clerics as by scientists.

Traditional habits of the mind dissolve slowly, but they do indeed deserve to be reviewed and revised in the twenty first century—not by dismissive mandates or government edicts; so much as a recognition that changes occur at different rates in different spheres of life. Some changes augur improvement; others seem to follow laws of entropy and decay. The task of the social sciences is measurement not so much of the worth of invention, but rather of the reasons why these rates vary substantially between the technical and the social. The vocation of publishing records such changes. If this is so, what are the variant relations between areas of human life? Leaders in communication have a golden opportunity to serve as bellwethers for this new paradigm because they have a foot in both the technical and social sides of the contemporary system. In this way the dogmas of the past will be buried, and the promise of the present, could become a source for future civilities. This in turn can lead to a greater appreciation of the democratic imagination, at least greater than then I understood so imperfectly in the past.

The notion of normative behavior resides in some measure on assumptions beyond the sociological or the anthropological. This is not to say that technical innovations do not automatically modify assumptions of the moral order. It is rather to say that human beings review such innovations slowly, haltingly, with political resistance and at times, personal resentment. Humans adapt to change on a need only basis. Within the bowels of scientific interpretation and constructive achievement are possibilities of terror—massive destruction on a scale that is unparallel. Five hundred years of Western Civilization should make it clear that change is no magic bullet for progress, nor for that matter are new formulas for social order policy mechanisms guarantees of such progress. We are left with decision-making in which mistakes in judgment are more lethal and total than in the past, so caution is the handmaiden of decent scientific method. It turns out that ethics is the durable ghost in the technical machine, but also limits the social whims of policy making and human engineering as well.

These remarks are not intended as a frontal assault against postmodernity and certainly not a critique of new information and communication technologies. The sort of Ludditism that hides behind moral absolutes as a means to lay at the foot of technology everything from identity theft to medial ailments that result from the use of cell phones has been repudiated too often to require further elaboration. What does need to be emphasized is how the very shift in technological models of growth can hugely impact everyday events.

The speed with which revolutions and rebellions, regicides and genocides is reported change the character and possible outcomes of a specific event. Instant reportage does not create the need for such measures as people will undertake in the name of democracy or in defiance of dictatorship, but they offer prospects of rapid moral judgment that tear away the obfuscation of those seeking to work and dominate behind closed doors. In short, while George Orwell's "big brother" watching the actions of ordinary citizens with the aid of advanced technologies does exist, so too does this technology just as readily turn its gaze on the big brothers. Communist officials in China order Google to cease providing access to overseas access in the Chinese language, in turn, Google is considering abandoning the Chinese market rather than submit to such political dictates. At the same time, the Iranian regime did everything possible to block access and usage to all Internet services during the post-election riots of June 2009, but to no avail. The information was recorded, photographed and sent forth worldwide. In commenting on the mass uprisings against the clerical dictatorship in Iran, Peggy Noonan caught the spirit of how moral judgments impacts technical discourse for the purpose of political change. Her remarks merit close attention. "The great question is what modern technology can do not in the short term as the long. It is not the friend of entrenched tyranny."

Thus, the answer to the opening query of Mortimer Kadish was implicit in the very context in which it was asked. It is one that I believe he would have wished us to acknowledge and appreciate. Yes, change is different in kind, space and substance for technology and society alike. But the history of ideas is not vanquished by executive decrees. This is why the irritating paradoxes of philosophy remain intact—as a useful preventative for correcting the hubristic impulses that infect the educated classes—in technology, politics and society alike. These hugely differential rates of change leave a space and a place for the humanistic purposes of the four inherited fields of philosophy: epistemology, logic, aesthetics, and above all, ethics—power points in their own right. Such differences in the rates of change in human life make possible better decisions in social policy and technical goals. Synthesis may remain a goal, but one of far less worth than dogmatic assertions from all quarters as to systems and even to the unity of science.

The plain and simple fact is that public policy is best when it navigates between technology and morality, and in recognizing that rabbits and turtles both have the right to live in the animal kingdom. When policy dramatically veers to one side or the other, the results are not synthesis, but chaos. In a democratic society the task of policy-makers and intellectual guardians alike is to make sure that these delicate balances of social forces are respected. The future will attend to resolutions of this grand sort.

References

Arendt, Hannah. (2000). *The Life of the Mind, Volume One: Thinking.* See the segment on "Banality and Conscience" in *The Portable Hannah Arendt*, edited by Peter Baehr. New York and London: Penguin Books.

Boguslaw, Robert. (1965). *The New Utopians: A Study of System Design and Social Change.* Englewood Cliffs, NJ: Prentice Hall.

Deresiewicz, William. (2009, December 11). "Faux Friendship," *The Chronicle Review/ The Chronicle of Higher Education,* (Section B): 6-10.

Duncan, Hugh Dalziel. (1962). *Communication and Social Order.* New York: Bedminster Press.

Gerbner, George. (1969). *The Analysis of Communication Content; Developments in Scientific Theories and Computer Techniques.* New York: John Wiley.

Gottmann, Jean. (1961). *Megalopolis: The Urbanized Northeastern Seaboard of the United States*. Cambridge, Massachusetts: MIT Press.

Horowitz, Irving Louis. (1969). "Engineering and Sociological Perspectives on Development: Interdisciplinary Constraints in Social Forecasting," *International Social Science Journal*. (UNESCO), (Volume XXI, Number 4): 545-556.

Katz, James E. (1994). "Values, Technology, and Administration: The Weberian Inheritance," in *The Democratic Imagination: Dialogues with the Work of Irving Louis Horowitz*, ed. by Ray C. Rist. New Brunswick, NJ and London, Transaction Publishers, 9-38.

Lasswell, Harold D. (1941). *Democracy Through Public Opinion*. Menasha, Wisconsin: Banta Publishing Company.

Mumford, Lewis. (1966). *The Myth of the Machine: Technics and Human Development*. New York: Harcourt, Brace & World.

Noonan, Peggy. (2009). "Whose Side Are We On? You Have to Ask?" *The Wall Street Journal*. (June 20-21, 2009): A13.

Pool, Ithiel de Sola. (1983). *Forecasting the Telephone a Retrospective Technology Assessment* Norwood, N.J.: Ablex Publishers.

Shen, David. (ed.). (2009). "The Next Step in Communication," *Technology First: Journal IV*.

Speier, Hans. (1989). *The Truth in Hell and Other Essays on Politics and Culture, 1935-1987*. New York: Oxford University Press.

19

Conclusion

James E. Katz

The preceding chapters have analyzed some important ways that the availability of mobile communication technology affects social policy as both a process and an outcome. They have also in some looked at social policies surrounding mobile communication technology. The levels that have been discussed have ranged for the psychological and sociological to the political and ideological. Questions of administrative structures and social interaction have been examined. The issues addressed include social program administration, community information systems and educational activities. Some authors have also engaged concerns over social policies for mobile communication use and availability. At this point, it is appropriate to highlight a few of the many crosscutting dimensions.

The first of these is that mobile communication devices have double-edged consequences for the policy process in that they can have both positive and negative aspects. This statement is of course not shocking but it is worth considering in terms of its implications. The positives include that they are in some cases able to ameliorate or resolve problems, and thus can be important adjuncts to social policy. Yet they can also create and exacerbate problems.

In terms of amelioration or resolution, the mobile device can do much to help people gain personal security. It can help them control their local environment, and the people within those environments for whom they have responsibility. Mobile devices can assist people as they plan for and seek to influence their futures. Likewise, mobile devices can be, at the individual level, important tools for managing people's health problems as well as getting in contact with healthcare professionals. On a targeted or nation-wide basis, they can form the basis of major health and public service campaigns and potentially deliver a variety of social services on a broad scale. The list of uses is in practical terms endless.

On the other hand, ubiquitous mobile communication can exacerbate both policy processes and other social problems. For example, the very ease of notifying the public about issues can prompt greater involvement from competing factions. This raises problems in either of two directions. One of these directions is that the mobile phone can be used to capture the issue by an organized and dedicated minority. This has already happened in online setting, but is equally applicable to the mobile setting. Specifically, in 2009 President Obama held his first "electronic town hall meeting." For that event, he said he would answer whatever question one the most number of votes in the run-up to the meeting. A coterie of dedicated activists organized themselves to generate the most votes for the politically embarrassing question of "Will you legalize marijuana?" Although the question did embarrass him, President Obama did address it because he had promised to do so, but he clearly would have preferred to have talked about his economic recovery plans (Katz, 2009).

The other direction is to tempt authorities to generate large numbers of notices. One result is that useful messages will become filtered out. In essence, the overuse of the service can result in its becoming of little or no use. In a perverse way, the reverse is also true. The sheer volume of reports that the populace can make can easily overwhelm organizational routines of authorities and the institutional capacity to respond.

Interestingly, it was feared that mobile communication would open elected officials to pestering by their constituents. The reasoning was that with the easy access of "anyone, anywhere, any time" which was facilitated by the mobile phone, officials would become constantly in touch with their constituents and colleagues. However, at least insofar as preliminary research undertaken by Leopoldina Fortunati (Personal communication) has shown, this is not the case. Moreover, email messages sent to members of Congress, once thought to be an ideal and powerful way for constituents to contact their elected representatives, has become largely ignored those offices (as opposed to email messages to specific staff) by governmental representatives, according to Tobin L. Smith of the Association of American Universities (Personal communication). So apparently, as has been the case of so many other technologies, including the fax, hopes for the direct exercise of democratic power-at-a-distance over elected representatives via telecommunication technology has not materialized.

In a world of accountability, and where there is a constant drive for "zero defects," such a tidal wave of unaddressed complaints represent a threat to the authorities and bureaucracies that claim to have responsibility for the area. In quite a different context, namely Puritan society of New England 400 years ago, Kai Erikson analyzed how too much deviance could overwhelm systems of justice (Erikson, 2002). Although we are far distant from that time, the simple truth remains that most social service organizations have no way of totally resolving the problems that they were formed to address.

Yet the availability and frequent use of the mobile also adds complications to the policy process arena itself. It allows greater opportunities for participation, but this makes the process far more complex and difficult to manage. It also allows greater opportunities for direct democracy. While such a prospect is attractive in nations that have traditionally not been democratic, or even have meaningfully engaged its populace rituals of legitimacy, it is not necessarily ideal in practice. Perhaps recognizing this fact, most governments have gravitated towards representative models. This is the case even though, with the near-universal possession of the mobile by every citizen, direct consultation on every governmental issue with the populace is now practical.

After all, with a mobile phone, voters could be consulted on every issue that comes before any governing body. This idea is not entirely far-fetched. In the 1980s, in Austin Texas, a "tele-democracy" party was formed for city council. The platform of this party was simple: every issue that came before the city council would be referred to the citizenry. Using their telephones (in those days there were no mobile phones, but the landlines would have been sufficient), the citizens of Austin could vote on each issue and turn. These might range from the licensing of dogs to declaring the city a nuclear free zone. However, the lure of full democracy was not sufficient to put any of the candidates into office. The larger point here is that most people prefer representative democracy most of the time. Although the chapter by Campbell and Kwak show the importance of mobile communication in the lives of educated citizenry, and other chapters have highlighted the role of the mobile in political campaigns, it is interesting that very little effort is being devoted to developing mobile phone applications that allow direct input to governmental officials.

Returning to the theme of capacity of governmental policy organs, it is possible to see how mobile communication can complicate administrative practices. The case of the Chinese earthquake highlights the problem of responsiveness. Once there is a way that authorities can be notified of a problem, then there is the demand that they address it. Yet as this incident shows, the capacities of many bureaucracies are overwhelmed when the situation they supposedly have prepared for actually strikes. This is seen not only in the Chinese earthquake response, but in the response to the

Katrina hurricane that struck New Orleans in 2003. As well sometimes the ability to communicate outstrips the ability to respond, even if the intervention bodies are prepared, due to distance and tenuous logistical lines, as was the case in Haiti in 2010.

Linking the above points of the limits on administrative capacity and public's disinclination to undertake fully democratic participation, we see that representative democracy will likely be enhanced due to mobile communication in terms of breadth of process involvement, but constricted in terms of efficiency. In terms of delivery of social services, the future is bright. Technology will make the delivery of services faster, cheaper, and more efficient.

In terms of social grouping, the idea of near-perpetual contact is likely to continue to grow. Many new services are becoming available, and that will be available on popular mobile devices such as the Apple iPhone. "The idea is to allow groups of friends or colleagues attending such events as a concert, a tradeshow, business meeting, wedding or rally to stay in communication with each other as a group to share information or reactions to live events as they're occurring" (Wortham, 2010). While some sober and reasonable scholars will see this constant chattering, and tweeting, as deleterious to the inner reflection that makes life worthwhile, if not bizarrely unattractive to any reasonable human, it is also possible to see that such activities can actually form a bulwark against religious or political extremism. The value of multiple loyalties has been shown to be a strong counter-balance to totalitarian ideologies of all stripes (Lipset, 1956). With mobile communication, a new way to have interstitial connection grows. Certainly such connections can be monitored and aggregated by remote authorities, but the sheer magnitude and metamorphosis of context makes the likelihood of the Panopticon world of centralized control much more remote.

At a larger level, the structures and processes of policy institutions tend to lag and are generally poor at incorporating the technology into their routines. This is marked contrast to businesses and the world of social interaction, which tend to be remarkably speedy adopters, and assimilators, of mobile communication technology. The difficulty that state governments have been confronting information policy in the use of mobiles, as discussed by Scott and colleagues, our case in point.

Crosscutting Issues

Could the mobile device be the most democratizing technology since the printing press? In many ways, it does for the ordinary person who has a working mobile what the printing press did for the person who in an earlier era assuming that that person could read and write, and somehow had access to the sundry fruits of the printing press. These fruits would have included broadsides, music scores and lyrics, advertisements, newspapers, political, and religious tracts and, of course, books. Of course these fruits were limited by those who in turn were able to get their material produced via a printing press. The barriers are much lower with the mobile device than with the printing press, of course, even as the mobile becomes a printing press of sorts.

Yet I also mean this in the sense of owning a printing press: mobile device often provides the individual with the ability to produce material and disseminate it.

We can look at the mobile phone itself, which is a subset of a mobile device, from the perspective of power, people and control. In terms of power, it allows people to have expanded control over their personal environment. It allows them to find others, and organize activities with those people. It allows them to summon various resources and transfer them to the hands of others.

The possibility of widespread mobile banking is significant in this regard because it does put control over resources literally in the palm of the user's hand. And over time having the mobile device serve as an alternative to central commercial and governmental monetary systems is intriguing. It is already something of an alterative to the international financial system as prepaid cards, sent minutes, and balance transfers all become part of the resources at the command of ordinary people, especially those from developing countries.

Though this overview has only highlighted a few of the many dimensions of mobile communication and social policy, it should be sufficient to allow several conclusions. These are admittedly daring, and one should always be cautious about predicting the future. That said, they might include:

- Mobile will be increasingly important to social policy as the saturation mobile devices approximate 100 percent of the population.
- Mobile communication will be increasingly used as the capability of devices grows while, at the same time, the infrastructural capability of organizations learns to accommodate and respond to activities in the mobile domain.
- Mobile communication will continue gaining an important role in fighting governmental abuse and resisting totalitarianism. This was amply demonstrated in the 2010 Iran national elections and is continually being used to pry open the Gulag nation of North Korea (Choe, 2010).

- Contention over the proper uses of mobile devices is likely to subside as society adjusts to the norms that govern both proper "front stage" behavior and the incessant need to multitask; this will take place even as the harms of doing so, especially while driving, are increasingly recognized and sanctioned.
- Mobile communication will be increasingly seen as a "human right." In contrast to the Internet, where hundreds of millions of dollars have been spent trying to overcome the digital divide, it is likely that the mobile phone will not receive such specialized treatment. In part this is because of the biases of policy makers. It is also in part due to the fact that people find the mobile phone so enormously valuable in their lives, that they will marshal their own resources to get access to them, as Loudres Portus discussed in her chapter. That said, yesterday's luxuries tend to become today's necessities. That was the case with the telephone itself. Originally in the United States the wireline telephone was considered a luxury. However, by 1968, it was considered a necessity and was included in the basket of welfare benefits that, at least in New York City, was given to poor people.

Given the entrepreneurial quality of policy enthusiasts, there will be a continuation of pressure to correct a perceived "digital divide" in access to mobile communication resources. While much of the Zeitgeist over digital divides originally settled on the Internet itself, then broadband access to the Internet, we can expect the focus to soon settle on mobile communication and eventually high-speed mobile communication access. This is understandable, especially given the vital role that the mobile phone can play in the economic lives of those who are at the lower rungs of society, and most especially those in such positions in developing countries. This is all the more the case given that mobile communication services consume such a vast proportion of the wealth of the poor in developing countries. For instance, BonTempo (2010) states that in Kenya, the bottom three-quarters of earners spend more than 25 percent of their income on mobile services

We can see that mobile communication can be important at every phase of the social policy process. Typically, the policy process itself is considered to have the following elements: issue identification, agenda setting, consideration of alternatives and persuasion to accept one such alternative, procedural steps to implement the planned policy, plan development and application of the policy, administration of the policy, evaluation of the policy and its impacts. As the chapters in this volume demonstrate, mobiles can affect each of these phases.

Considering social policy, it is also important to view the process from the perspective of the people who will be affected by the policies. More particularly, considering the perspective of someone who would wish to receive social service, mobiles can play role in alerting them to the possibility of the service being available, being the modality of access to the service, and even the platform for the delivery of the service. This could for example be the case with a retired person who would find out about a government benefit, such as money being available, and could securely apply for the benefit through the identity verification offered by a mobile. The benefit itself, in the form of money, could be delivered as credit to that individual's mobile. And the mobile could provide a way to provide feedback to the administering agency about the quality of its service.

Hyperbole seems to typify our age, perhaps as much as the mobile phone. Bold claims for the importance of this or that idea or technology too often typify public discourse. For example, it has been claimed that the Internet is the greatest invention since the wheel or the printing press (these claims are reviewed in Katz and Rice, 2000, p. 29), and even I have been tempted to make them, as per the paragraphs above. Yet despite my inherent disinclination to claim too much, I cannot help to wonder just how revolutionary a technology is the mobile communication device. Especially in terms of social policy, it does seem that the mobile phone is able to do much of what the printing press did and does in terms of communication, especially of sending messages from a central location to a mass audience. And, of course it does what the landline telephone did, except in an often easier and more compact fashion. Too, in the sense of accomplishing with a message what would have required a personal journey or physical distribution of content, it does some of what the wheel did for human transportation. Finally, the Internet and the mobile phone are increasingly co-mingling those aspects that each had formerly done distinctively.

While it is tempting to take this line of argument too far, there seems little question that mobile communication is becoming a transformative force in social policy as much as in other spheres of human affairs. It fosters greater accountability in policy issue definition, processes and outcomes. It allows more finely nuanced delivery of social program content as well as more detailed analysis of behavior at the level of both the staff that delivers and oversees the programs as well as the people who are the subjects of the program. The Weberian delineation and fearful prediction as to the ultimate evolution of the iron cage of bureaucracy is strengthened by a web of information and communication enabled by the mobile phone, and decision-making becomes more inclusive but also more complex and cumbersome.

This results in an ironic intermingling of greater freedom and increased responsibility and accountability for people. Privacy as a valued concept recedes as the norm of exposing personal details becomes more prevalent, and in some sense irresistible as the technology of communication and life-tracking become omnipresent. The long-term implications for civility are troubling as the polite fictions of life come under attack (Becker, 1969). Yet it is also the case that our moral vocabulary has not kept with technologically driven practices. The terms of "privacy" and "solitude," while not outmoded, and remain vital reservoirs for the protection of individuality and liberty, do not fit well with much of contemporary youth culture. Legal and educational frameworks need to evolve to address the race, as described by Horowitz in the preceding chapter, between the technological rabbits and communication turtles. The temptation to see communication practices not from the preferred stance of value neutrality, which has been the hallmark of social scientific inquiry, but rather from moral perspective of decay, decline, and pollution is often irresistible. Yet there are ever-renewing generations who wish to create the world in their own image. Mobile communication as the tool, and social policy as the arena, is allowing them to do so on a daily basis. The work of billions of mobile-enabled humans via this new tool will re-make the world. Recent evidence is mixed as to the likely trajectory on many levels, but in terms of fostering democracy, transparency, and public engagement, the future holds tremendously exciting possibilities.

References

Becker, Ernest. (1969). *Angel in Armor*. New York: George Braziller.

BonTempo, James. 2010, (March 28). Cell phones, ubiquitous in the developing world, can be used to save countless lives. Baltimore Sun. Retrieved March 29, 2010 from http://articles.baltimoresun.com/2010-03-28/news/bal-op.cellphone0328_1_mobile-technology-mobile-phones-cell-phones.

Castells, Manuel, Mirela Fernandez-Ardevol, Qiu, Jack Linchuan, and Arba Sey (2007). *Mobile communication and society: A global perspective*. MIT Press: Cambridge, MA.

Choe, SANG-HUN. (2010 March 28). North Koreans Use Cellphones to Bare Secrets. *New York Times*. Retrieved April 1, 2010 from http://www.nytimes.com/2010/03/29/world/asia/29news.html?hp.

Erikson, Kai (2002). *Wayward Puritans: A study in the sociology of deviance*. Boston: Allyn & Bacon.

Katz, James E. (2009). Media, la democrazia, e l'Amministrazione Obama: la speranza, senza cambiare? *Politica Communicazione, 10* (3), 421-31.

Katz, James E. & Ronald E. Rice. (2002). *Social consequences of Internet use: Access, involvement and expression*. Cambridge, MA: MIT Press.

Lipset, Seymour Martin, Martin Trow, and James S. Coleman. (1956). *Union Democracy: The Internal Politics of the International Typographical Union*. Glencoe, IL: Free Press.

Wortham, Jenna (2010, March 18). "Is Apple throwing its hat into the location ring?" *New York Times*, March 18, 2010).

About the Editor and Authors

About the Editor

James E. Katz is the director of the Center for Mobile Communication Studies at Rutgers University where he is also the chair of the Department of Communication. He holds two patents in the field of telecommunication technology and is a fellow of the American Association for the Advancement of Science (AAAS) and the Foreign Policy Association. His books include *Magic in the Air* and *Connections*, both published by Transaction.

About the Authors

Julie Amoroso was an undergraduate research assistant at NetLab, and has recently completed a master's degree in policy studies at Queen's University.

Lora Appel is a master's student in the communication program at Rutgers University, and a research assistant to Professor Katz. She also holds a fellowship at Ethicon, Johnson & Johnson. She graduated with an international bachelor of business administration from the Schulich School of Business in Toronto, Canada, and furthered her studies earning a postgraduate certificate in corporate communication at Seneca College. Lora intends to advance her career in communications in the health industry and new media, and is interested in pursuing a PhD.

Nisar Bashir holds the position of researcher in Telenor Group Business Development & Research at the corporate headquarters in Oslo, Norway, where he has worked since 2006. His current focus is the evolution of payment systems to meet mobile Internet demands. Nisar holds a B.Sc. Hons degree in electrical engineering from Queen's University, Kingston, Canada.

Gillian Bonanno is a doctoral student and graduate assistant in the School of Communication and Information at Rutgers University. She received her master's degree in corporate communication from the City University of New York, Baruch College, where she was named to the Sigma Alpha Delta Honor Society. Ms. Bonanno was employed as a training specialist and implementation manager for over nine years in the financial and retail industries.

Scott W. Campbell, Ph.D. is assistant professor of communication studies and Pohs Fellow of Telecommunications at the University of Michigan. Professor Campbell's research explores the social implications of new media, with an emphasis on mobile telephony. His current projects examine how mobile communication patterns are linked to both the private and public spheres of social life, such as social networking and civic engagement. Several of these projects use a comparative approach to situate the role of mobile communication technology in the larger media landscape.

Jonathan Donner is a researcher in the Technology for Emerging Markets Group at Microsoft Research India. His primary research interests concern the economic and social implications of the spread of mobile telephony in the developing world. Previously, he was a post-doctoral research fellow at the Earth Institute at Columbia University, and worked with Monitor Company and The OTF Group, consultancies in Boston, MA. His Ph.D. in communication research is from Stanford University.

Leopoldina Fortunati is professor of sociology of communication at the University of Udine (Italy), Department of Economics, Society and Territory. She researches cultural processes (fashion) and communication and information technologies (especially fixed telephone, mobile, Internet, online newspapers and print press). She is active at cross-European level, and especially in COST networks. Among her publications is the co-edited (with Jane Vincent) volume, *Electronic Emotion. The Mediation of Emotion via Information and Communication Technologies* (Oxford, 2009).

Melissa E. Fritz received her M.A. from Carleton University in mass communication, and PhD from the University of Toronto, Faculty of Information. Dr. Fritz is interested in research for policy analysis. Her research has focused on issues of digital divide, telecommunication policy, gender and technology adoption, and motherhood and child care in Canada. She is continuing her research on broadband deployment in rural and remote areas, emphasizing gender and information practices as they relate to e-health initiatives.

Gerard Goggin is professor of digital communication and deputy director of the Journalism and Media Research Centre, University of New

South Wales, Sydney, Australia. He holds degrees from the Universities of Sydney (PhD in English literature) and Melbourne (B.A. Hons). Dr. Goggin's research interests lie in social, cultural, and policy aspects of new media and communication technologies, especially mobiles and Internet, and also in disability. His books include *Global Mobile Media* (2010), *Cell Phone Culture* (2006), and, with the late Christopher Newell, *Disability in Australia* (2005), and *Digital Disability* (2003).

Gordon A. Gow is associate professor in communication and technology at the University of Alberta. He examines the impact of new communication technology on public health and safety. His areas of expertise include mobile and wireless communication, telecommunication policy, Internet and social media. His recent work has included studies into the development of next generation public alerting systems in Canada as well as community based hazard warning and health communication projects in Sri Lanka and India. Dr. Gow is also director of the MARS micro-lab (Mobile Applications for Research Support) at the University of Alberta and provides support for projects investigating the use of the mobile communication devices in support of scholarly and community-based social science research. From 2003-2006 Dr. Gow was a lecturer in the Department of Media and Communications at the London School of Economics and Political Science, where he was director of the Graduate Programme in Media and Communications Regulation and Policy.

Tracy L. M. Kennedy is a PhD candidate in sociology at the University of Toronto. Her doctoral thesis investigates how Canadian households have incorporated ICTs into their homes and daily routines. She teaches part-time in the Department of Communication, Popular Culture & Film at Brock University on topics such as digital culture and social media.

John Leslie King is William Warner Bishop Collegiate Professor of Information and vice provost for academic information for the University of Michigan. Dr. King joined SI on January 1, 2000 as professor and dean of the School. Previously, he was professor of information and computer science and management and a research scientist in the Center for Research on Information Technology and Organizations at the University of California, Irvine. He conducts research on the development of high-level requirements for information systems design and implementation. Drawing on engineering and the social sciences, he studies the organizational and institutional forces that affect how information technology is developed.

Nojin Kwak, Ph.D. is associate professor of communication studies at the University of Michigan and a director of the Center for Korean Studies. Professor Kwak's main research centers around the influence of mediated and interpersonal communication on political and civic participation, knowledge, and attitudes. One of his current research projects focuses on the role of nontraditional channels of political communication, including soft news, new communication technology, and political talk.

Chih-Hui Lai is a doctoral student in the Department of Communication, Rutgers University. Her research focuses on how individuals, groups and organizations use information and communication technologies (ICTs) to communicate and how relationships evolve or emerge through the process. She analyzes social implications of online and mobile applications and examines how a social network perspective can illuminate the communication patterns among users.

Sun Sun Lim (PhD, LSE) is assistant professor at the Communications and New Media Programme, National University of Singapore. She studies technology domestication and charts media ethnographies in Asia, having conducted research in China, Japan, South Korea and Singapore. She has articles published the *Journal of Computer Mediated Communication*, *New Media & Society, Telematics & Informatics*, and *Science, Technology and Society*. She also sits on Singapore's Internet and Media Advisory Committee and the National Youth Council.

Rich Ling is a professor at IT University of Copenhagen and a researcher at Telenor. He received his Ph.D. from the University of Colorado in 1984. He has also been the Pohs visiting professor of communication studies at the University of Michigan in Ann Arbor, Michigan and now he holds an adjunct position in that department. He is the author of the book: *New Tech, New Ties: How Mobile Communication Is Reshaping Social Cohesion* (winner of the 2009 Goffman Award) and *The Mobile Connection: The Cell Phone's Impact on Society*. He has been interviewed on *The Discovery Channel, National Public radio, MSNBC* and Norwegian TV, among other outlets.

Anna Maria Manganelli is full professor of social psychology at University of Padua, Department of Applied Psychology. She works in the fields of communication and information technologies, research methodology, social psychology (theories of prejudice and measurement of "modern" forms of prejudice and sexism) and economic psychology (test of models in the formation of intention to perform economic behavior). Recently she published in national and international journals—*Journal of Economic Psychology* and *Personality and Individual Differences*.

Rhonda N. McEwen is an assistant professor at the Faculty of Information, University of Toronto. She holds an MBA in information technology from City University in London, England, an MSc in telecommunications from the University of Colorado, and a PhD in information from the University of Toronto (2010). Dr. McEwen's research and teaching centers around information practices involving new media infrastructures, with an emphasis on youth media literacy, mobile communication, and social media design. Her article, "Tools of the Trade: Drugs, Law, and Mobile Phones in Canada", was published in *New Media & Society*.

Amanda McGarry is pursuing a master's degree in vocal performance at Rutgers University where she also assists James Katz. As well, she serves as assistant editor for *Human Communication Research*, a leading journal in the field. She graduated *summa cum laude* from New York University with a bachelor's of music in vocal performance, specializing in classical music. Her interests include the interpretation of song, the pursuit of communicative clarity in performance, and text messaging.

Lourdes M. Portus is an associate professor and concurrently the college secretary of the College of Mass Communication (CMC), University of the Philippines (UP). She obtained her Ph.D. in communication, MA in communication research and AB in social sciences from the University of the Philippines. She has published a book, *Streetwalkers of Cubao* and co-edited another book, *Building Social ASEAN*, among her publications. Her varied interests include ICT studies, health risk communication, communication research, training, project development and management, and monitoring and evaluation.

Craig R. Scott is an associate professor of communication and director of the Ph.D. program in the School of Communication and Information at Rutgers University. He received his Ph.D. from Arizona State University. His research examines communication technology use, identification, and anonymous communication in the workplace. His work related to communication technology use by organizational members. Dr. Scott has been widely published in outlets such as *Communication Monographs,*

Communication Research, Management Communication Quarterly, and *Journal of Computer-Mediated Communication.*

Matteo Tarantino, Ph.D., is a lecturer and researcher at the Università Cattolica del Sacro Cuore of Milan. His research interests focus on the sociology of the media and technology. He has collaborated with the Communication University of Beijing and the Hong Kong Polytechnic University.

Minu Thomas completed her MA at the Communications and New Media Programme, National University of Singapore. She is currently working in India as an education services professional.

Nuwan Waidyanatha is a senior researcher at LIRNEasia, a regional ICT policy and regulation think-tank active across the Asia Pacific. He recently served as director for LIRNEasia's mobile health project: Evaluating a Real-Time Biosurveillance Program as well as being a lead Researcher/Project Manager of the disaster project Evaluating Last-Mile Hazard Information Dissemination in Sri Lanka (HazInfo). He is also a member of the Project Management Committee of the Sahana Disaster Management Framework Free and Open Source Community playing a key role in the developments on a public Messaging/Alerting system. Nuwan received his bachelor's degree in computer engineering and mathematics from University of Maine (1995) and master's in operations research from University of Montana (1999).

Marion Walton is a senior lecturer in the Centre for Film and Media Studies at the University of Cape Town, South Africa. Her PhD studies (Computer Science, UCT), also included a period of study at the Centre for the Study of Children, Youth and Media in the Institute of Education, University of London. Her research in human computer interaction approaches the study of software as a new form of media, and confronts the issues of power and regulation of meaning that arise for users of software, particularly those in marginalized contexts.

Barry Wellman is the S.D. Clark Professor of Sociology at the University of Toronto where he directs NetLab—a research network at the University of Toronto that studies the intersection of computer, communication and communication networks. A member of the Royal Society of Canada, Wellman is the (co-)author of more than 200 articles and the (co-)editor of three books. He is currently writing *Networked* with Lee Rainie (MIT Press, 2011).

Yun Xia is an associate professor in the Department of Communication and Journalism at Rider University (USA). He received his Ph.D. from Southern Illinois University. His research interests are on the social impact of computer-mediated communication, educational applications of communication technologies, visual intelligence of graphic communication in new media, and semiotic analysis of communication signs in new media. His works have appeared in journals such as *Human Communication, China Media Research,* and *The American Journal of Semiotics.*

Hyunsook Youn is pursuing her doctoral degree at the School of Communication and Information at Rutgers University. Her research interest is focused on use of communication technology and its impact on social relationships. She is also interested in examining individuals' concerns about online privacy issues and individual's use of communication technology regarding health-related issues. She finished her master's degree at Florida State University and her work has been published in *CyberPsychology & Behavior* and *Health Information on the Internet.*

Index